CARAVANS
The Illustrated History from 1960

DEDICATION

Dedicated to my parents Jack (1918-2001) and Eileen (1918-1979) both sadly missed, and always in my thoughts. Also to my lovely daughter, Georgina Olivia, who has given me more joy and love than I could ever have imagined and is an inspiration in many ways.

First published in 1999 by Veloce Publishing, 33 Trinity Street, Dorchester DT1 1TT, England. Paperback edition published 2003.
Fax 01305 268864/e-mail info@veloce.co.uk/web www.veloce.co.uk or www.velocebooks.com
ISBN 1-903706-85-8/UPC 36847-00285-5

Andrew Jenkinson also produces a range of classic caravan scene postcards and blank greeting cards. For more information log onto the website - www.jenkinsons-caravan-world.com

Readers with ideas for automotive books, or books on other transport or related hobby subjects, are invited to write to the editorial director of Veloce Publishing at the above address.

British Library Cataloguing in Publication Data -
A catalogue record for this book is available from the British Library.

Typesetting (URWImperialT), design and page make-up all by Veloce on Apple Mac.
Printed in the United Kingdom.

CARAVANS

The Illustrated History from 1960

Andrew Jenkinson

Foreword

Following on from my first volume (1919-1959) I continue the story of the British touring caravan industry as it expanded at a faster rate than at any time previously. New manufacturers appeared, such as Swift, Ace, Elddis and Avondale, names now well known in caravanning circles. The Hull area witnessed most activity as it became the centre of caravan production.

With the touring caravan becoming ever more popular, by the mid-60s Sprite had become almost a household name. Caravans International, which was founded with the Bluebird/Sprite merger, not only promoted tourers, but caravanning as a whole. The company also bought caravan component makers and so provided a CI spares back-up system, even for those users on the continent.

Ace Caravans had become a formidable force and Thomson was still growing and expanding export markets. New dealers came and went, while some grew to become the large concerns they are today. In the 70s, the British caravan industry was in full swing with large export markets. With the downturn of business in the early 80s, those who had invested in new machinery and training expanded and cornered new markets.

Touring caravans also became better equipped, with electricity, hot water, full insulation and double glazing, whilst also being lighter and more stable on tow.

In recent years, tourers have been given a bad press by the media. I hope those who have contributed to this negative approach will read this book and realize that this is truly one of the last UK industries.

I hope you enjoy reading this book as much as I have enjoyed writing it, and that through my never-ending quest for archive material, it gives you an insight into how the touring caravan has developed into the sophisticated leisure vehicle of the 90s and beyond.

Contents

Sam Alper's Caravans International

As the 1950s ended, the caravan industry began to develop at a faster rate of knots than any other time in its history. From its early beginnings, progress in caravan manufacture had been evolutionary rather than revolutionary. Only Eccles had shown major advances in both design and production techniques. This was all about to change. In this first chapter we cover the rapid growth of Sam Alper's caravan manufacturing empire. From 1960 up to the early 70s, the company, Alperson Products (Sprite), became a global name within this sector of the leisure industry.

Alper strove hard to provide value for money caravans and he also actively sought new markets. His ability to raise his own company's profile, and that of caravanning in general, made Alper's name within the industry. His company's rapid growth had seen annual Sprite production increase from under 1000 units in 1952 to over 2500 by 1957.

The names that Alper gave his models, such as Alpine, Major, Colt and Musketeer, for many years became synonymous with touring caravans. At the end of 1960, Sprite production had topped 5200 units. This figure was up a full 37% on 1959. Alper's caravans were distributed in 17 different countries and were being manufactured in South Africa, Rhodesia and the Republic of Ireland. For a short time, the Sprite 14 of 1956 was being assembled in France by Digue.

Thanks to heavy investment, production at the company's headquarters at Newmarket in Suffolk had been able to reach new targets. New paint shops had been installed and now all stages of manufacture were being accomplished at the Newmarket plant. Gone were the days when Alper had to subcontract work to other firms. One problem he did face, though, was the fact that, during the winter months, work at the factory tended to drop off. This meant that his workforce had to be reduced, resulting in a shortage of labour when the time came for production to be increased again.

Alper came up with the idea of producing a football board game for children, (Soccerette) which meant that he could continue to employ

The Newmarket plant in the early 60s was a hive of activity. Expansion was a word often used by Alper; note how ideally suited it is for such purposes. (Courtesy Dave King)

Beginning production of the Sprite 400. Rolls of glass fibre insulation are all ready for use. (Courtesy Dave King)

his staff even in slack times. He did this for around 8 years and, although profits were small, it paid off through not having to try and re-employ staff for the new season. Even Alper's top designer, Reg Dean, was recruited to use his illustration skills to design the packaging for Soccerette and other games which the Sprite factory produced. The company not only made board games but, for a time, diversified into making small sailing craft, calling them Sprite dinghies. Caravans were the main concern, however, and, although the company concentrated mainly on tourers, it also built static vans such as the Elizabethan, Fenman and Escort.

Alper's chance to expand came about with the first of his takeovers. Eccles, which had been founded by the Rileys in the Midlands, produced good caravans and was the oldest name in the industry. Owner Bill Riley was now looking at selling the company (tragically his only son had died of cancer in his 30s). Although still a name with credibility, Eccles had been overtaken by some of the newer companies in the industry. Alper saw his opportunity and, in August 1960, purchased the Eccles name.

Some thought this takeover would spell disaster for the Eccles reputation, but, in fact, Alper had the time, energy and finance to invest in the company. Within two years, Alper moved all production from the Midlands down to the Sprite factory in Newmarket. Alper made the Eccles brand a priority and set about redeveloping the range. For the 1962 season a new range of Eccles vans was launched, to be known as the "Newmarket Eccles" because of their modernistic design, both inside and out.

The new Eccles came from the drawing board of Reg Dean, a designer who became the most prolific within the caravan industry over the next 30 years. Dean, who had been a freelance furniture designer and illustrator, lived in the town of Newmarket, not far from the Sprite works. In 1957 he decided to look for work in London. Alper, in the meantime, had decided that tourer design was becoming more specialised. He decided to put an advert in the Industrial Design Centre. Dean came along, saw the advert for a caravan interior designer, applied and got the job.

Although he had never even looked at caravan design, Dean's move into the industry in 1957 was to become his life. His new approach to caravan interiors was a breath of fresh air, and his reputation within the trade grew to such an extent that he was widely headhunted.

Framework done, the insulation was added and the sides went on. (Courtesy Dave King)

Sprite 400 production in full swing. Fittings are kept alongside the completed shells. The models under construction here are all for export. (Courtesy Dave King)

Dean's style was very obvious in the new-look Eccles. Teak-effect furniture combined with white vinyl covered walls made the vans' interiors feel modern and spacious. Exteriors took on a bay front window and sloping ends, which, besides looking up to date, also made the brand more distinctive. Model names such as Sapphire, Moonstone and the GT 305 also added to the Eccles' change of image. Dean was to leave CI early in 1965 to join Manchester-based Lynton, where he made an even greater impact on the company's fortunes and status.

Alper was now producing nine models in his Sprite range. These included the new 400 Mk1 in 1960, a 10ft four berth, and the 38ft Santa Monica mobile home. Alpine, Musketeer and Major were other long standing names in the Sprite tourer line-up. Sprite's involvement with mobile homes and static holiday vans decreased as the 1960s progressed, and had disappeared completely from the Sprite line-up by 1969. The CI motorised division (later CI Autohomes), maker of the well known Bluebird Highwayman, also made motorhomes with the Sprite badge associating the Sprite name with motorhomes as well as tourers, holiday and mobile homes.

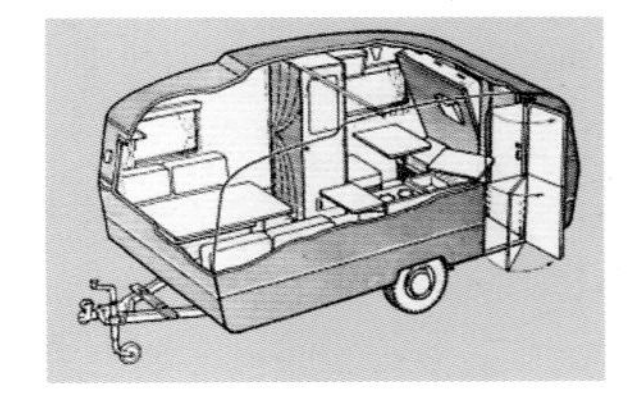

Sam Alper was constantly striving for increased production and export opportunities, which were always to play a large part in Sprite's success. Problems, though, were never far away. On June 25th 1961, disaster struck in the form of a serious fire in the paint shop. Production was hit by a 2% cut in manufactured units, compared with 1960. This latest setback occurred after the company had just overcome financial problems brought on by over-expansion due to high demand.

Alper was granted time by his creditors, which enabled him to turn the company around by reorganising management and the executive structure. This revamp of the Sprite management team gave the opportunity to further strengthen the important backup aftersales service, thus resulting in better relations between customers and dealers.

Endurance tests were still playing a major publicity role in proving the durability of the Sprite, and these continued right up to the late 70s. In Holland, during the summer of 1962, for instance, a Sprite Musketeer and a 400 model were towed around the Zandvoort Grand Prix circuit. Over 1000 miles were covered in 24 hours, proving to continental buyers that the Sprite was a well designed tourer which could safely be towed at high speed. The Sprite's own chassis design and manufacture, plus the Sprite-designed suspension, was largely responsible for keeping it stable throughout the test.

Expansion plans at Newmarket were again being implemented. During 1963, the company erected a further 7000 sq ft building, making a total of 103,240 sq ft. Sprite and Eccles could now make over 10,000 tourers a year, of which 4700 were exported to Europe. Not only was Sprite establishing its own export markets, but was also, perhaps unknowingly, paving the way for the up and coming Hull makers, too.

Over in Poole, Bill Knott's Bluebird range covered just about every part of the caravan market, including the first Bluebird motorhomes in 1959. Knott had also decided it was time to float his company, which involved changing its name from Bluebird Caravans to Bluebird Investments.

Bluebird had grown considerably since it started in 1935. With his knack of buying bulk raw materials cheaply, Bill Knott produced competitively priced caravans, which put him into second place in caravan manufacture on a global scale. Investor confidence in Bluebird saw shares being issued at 14 shillings and 6 pence (72p), rising to 18 shillings (80p).

Bill Knott took a back seat in the company in later years before leaving and setting up BK Caravans (holiday homes) at Mannings Heath, not far from the Bluebird works. At the end of 1962, rumours were rife about the possible merger of Bluebird and Sprite. Alper and Knott

These shots are from around 1964/5. The days before pre-finished aluminium panels meant a paint shop in operation in every caravan factory. The model next to the 400 looks to be a Sprite Fenman. Tourers and holiday homes were built alongside each other in those days. (Courtesy Dave King)

knew that such a venture would give them domination of both home and continental markets.

Although it was thought that the talks between Bluebird and Sprite would eventually lead to a negative outcome, the reverse, in fact, was true. After what seemed like months of speculation, eventually, on the 29th of July 1963, Sprite and Bluebird Caravans merged. This led to the formation of Caravans International, or CI, as it became known. At last, this made the caravan industry a far more respected concern among other industrial giants, andmeant that CI had stronger buying power amongst specialist suppliers.

Bluebird was to concentrate more on holiday and mobile homes, though, for six years, tourers still carried the Bluebird logo. Tourer production continued at the Poole works, with new tourer ranges being developed. Bluebird's tourer line-up included vans such as the Wren, Bantam, Dauphine and the Europe. The Europe was the first tourer developed after the merger in early 1964. Aimed at the small car owner - and with more than a hint of continental style - the Europe was only 10ft in length. The other plus point was its unladen weight of just over 9cwt. Internal features were characterised by very clean lines, made possible by eliminating knobs and handles from the furniture.

Sprite's involvement resulted in new production ideas being adopted at Poole. Special departments were set up to deal with different stages of manufacture and introduction of stiffer quality control procedures. One of the most essential developments was a dedicated aftersales service. Sam Alper had found that customers wanted a good backup spares supply, a service which had been lacking in the industry in

Several Sprite export 400s are lined up outside the assembly plant ready for dispatch. Caravan production has changed very little - still relying on manual labour for much of the process. (Courtesy Dave King)

general (Alper had discovered this early on with his Sprites).

Bluebird launched its brand new three-van Europe series for 1965. Ranging from 9ft to 16ft in length, the new vans replaced all previous Bluebird tourers, including the long-running 16ft Sunparlor. Caravans International's influence on the new Europe tourers saw the three models given a full testing at the MIRA test track. It was at this test site that Sprite pioneered caravan road testing as far back as the mid-50s. The Europes were not put into production until CI engineers were satisfied with the vans' handling qualities.

One of Bluebird's most interesting tourers, produced just before the Sprite merger, was the Bluebird Joie de Vivre, an 18ft, six berth Luton top caravan with the door at the rear. Launched in 1963, and based very much on American designs, it was soon followed by Astral's Sunrise, which copied the Bluebird design to almost every curve. Both models had limited success and continued for only two seasons. Not long after CI had been formed, the company saw further growth when, in 1964, Sprite Pty in South Africa joined the group (which then went on to purchase Africaravans for around a quarter of a million pounds). Interestingly, Africaravans was partly owned by the Thomson family, who manufactured caravans in Scotland. In that same year talks were held with Wilk caravans in West Germany. Alper visited the works there and came away with the strong impression that Wilk could become a growth sector of CI.

Sprite's famous Alpine from 1965. The 12ft tourer was to prove the best selling caravan of all time.

Later in 1964 the Wilk company became the newest member to join the group. This gave CI a further share in export markets and, for a short spell in the 60s, Wilk tourers were offered to UK customers as well. This latest takeover by CI sent waves of fear through the West German caravan industry; so much so, that several German makers formed a caravan union, the idea being to stay independent but pool resources and research for aftersales and service, thus fighting off any more possible foreign takeovers.

Fairholme Products, the Cardiff tourer manufacturer, had, like Eccles, found its market being infiltrated by newer makers. Along with some bad decisions on design alterations, Jim Hennessy, Fairholme's founder, saw sales beginning to tail off. The company began to lose its lead, and even fall slightly behind the competition, for a number of reasons. For instance, it was the last quantity manufacturer to change over to fully independent suspension units, as supplied by B&B Trailers.

Hennessy had always been on friendly terms with Alper and both enjoyed a good competitive relationship. Alper had admired the Fairholme tourer, and was aware that the company still had a good reputation, although, at this time, he hadn't foreseen Fairholme having a major role to play in the CI group. Jim Hennessy, it is alleged, asked Sam Alper to take over the Fairholme company. Whatever the story, in 1965, the CI group did take control of Fairholme and began to make changes to the Fairholme range, introducing new models along with revamped exteriors. Production stayed at Cardiff until 1969, when the whole operation was moved to Newmarket. Fairholme now benefitted from new investment and, in later years became CI's upmarket clubman range. Hennessy was still very much in control, along with his right hand man, Colin Thomas, who was in charge of home and overseas sales.

Fairholme's early 60s tourers dropped the old model names of Babette, Bambino and Harlequin, in favour of the New Dawn models, with the lengths becoming the model identification system. As with Eccles, the CI machine went into action and gave the Fairholme tourers new exteriors. These were not as radical as the Eccles of a few years earlier.

Two factories were now operating at the Newmarket plant: the Oaks produced Sprites, while the Pines built the Eccles tourers. A few years later, the Bluebird Europes joined Sprite production while Fairholme moved alongside Eccles. Over 1000 people were employed between the two plants, making 35,000 tourers annually in the early 70s. When you compare that figure with the total UK output of 22,000 today, it's clear to see what a giant operation CI was (as well as the scale of the market for new caravans as a whole).

CI not only invested in new general machinery, but also used raw materials such as plastic in caravan construction. Eccles experimented with vacuum-formed sheets of plastic in various designs. Being innovative, Eccles began making items to use in its tourers: soap dishes, cutlery drawers and cup racks were all produced. With better machines and more experience, Eccles became more sophisticated in the use of plastic, and could, by the late 60s, make the famous Eccles front bay window panel in one seamless piece.

Eccles went further by producing plastic gas locker fronts and tops. Roof vents, another good use for plastic, were made not just for Eccles tourers, but also all the rest of the CI brands as well. In addition, Eccles used its vacuum-forming machines to make scenery for studios, and intricate props, such as regency carved doors, could be reproduced in no time.

Alper's caravan organisation was now growing so fast that, by 1966, both Eccles' and Sprite's combined efforts resulted in the company receiving the Queen's Award to Industry. This was a first in the caravan trade, and was soon followed by the Hull makers a few years later, an indication then of just how strong and quickly the industry had grown. The British Caravan Road Rally, an event briefly mentioned in my first book, was, to many caravan manufacturers, a showcase for how their caravans stood up to the rigorous and often tortuous punishment received over several days of rough towing. From hurling tourers around race tracks, the vans were towed down unmade roads, which reduced many to matchwood.

Being the largest manufacturer, CI was keen to support the event and entered Sprite and Eccles models in all the classes it could. In 1966 both dominated the rally, proving to the public and caravan press alike that CI was producing good quality tourers. CI used the rally as further publicity, alongside the endurance tests, satisfying Alper's appetite for exposure of the CI tourer ranges.

CI was to further develop the Eccles range

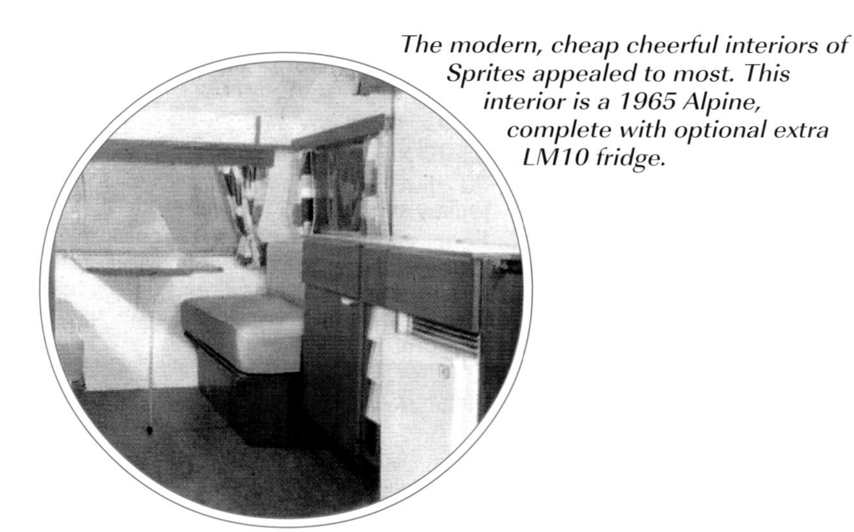

The modern, cheap cheerful interiors of Sprites appealed to most. This interior is a 1965 Alpine, complete with optional extra LM10 fridge.

with the introduction for 1967 of the 16ft, four berth Sapphire Winter Sports. Designed to meet the exacting demands of the winter caravanner, the Sapphire came with double-glazed windows, a heater and better insulation. At £595 it was a good £90 more expensive than the standard model and ran for three years. The Eccles name made news again in late 1969 with a new model, the Amethyst. With CI the size it was, Alper was able to use his organisation in a further attempt to promote the industry by offering more dealer support. Alper's long-term viewpoint led him, early in 1965, to offer 5 year loans for dealers to improve forecourt facilities and appearance. He was anxious, though, not to convert them to CI depots. In addition, to increase customer confidence in aftersales service, Sprite dealers could send staff to the Newmarket plant to be trained in the latest techniques in Sprite aftercare. Sprite tourers, by 1965, could take the Electrolux LM10 fridge as a standard works extra, and the drop-down double bed, as used in the Alpine and Musketeer T, was being abandoned in favour of bunks. Surprisingly, for a maker the size of Sprite, customers could still purchase a Sprite tourer in shell form, to fit out themselves.

From the late 50s, the small car market had been catered for by the Sprite Aerial, a name later revived in 1977. For 1960, the Sprite 400 replaced this model (called the 400 due to its weight of only 400kg and 10ft length). It became a best seller and ran until 1976. Nine years later it re-emerged for the 1985 model year, carrying on until 1986 with the same layout as its predecessor.

This is what the Sprite 400 looked like inside in 1965. These cost £265 new.

Such was Sprite's domination of the mass market it seemed that nothing could penetrate its stronghold at home or abroad. However, this didn't stop some of the newer makers trying to emulate Sprite models, such as the 400. Astral, who became a very big name in the industry, launched the 10ft Spree, which echoed the 400's profile in almost every way. Even the Thornton-based Summerdale had the 10ft Sport which, again, followed very closely the Sprite 400 lines.

The sheer size of CI meant its mass produced, big volume Sprite production could not be competed with, never mind matched or beaten. In the 1960s nobody was large enough to challenge Sprite's budget market leadership. That, though, would change in the early 70s. Bill Knott had, since Sam Alper's involvement, taken a back seat and eventually left the Bluebird company. Part of the agreement, between Alper and Knott was that after Knott left Bluebird, he would not be involved with caravan manufacture for several years. As soon as the agreed time was up, Knott returned to the industry with his new company, BK, producing holiday home caravans. These were largely built along the same lines as the Bluebird holiday caravans. Knott, though, wouldn't return to tourer manufacture, and leaves the CI story here.

As mentioned earlier, Eccles surprised everyone with the launch of a brand new model in late 1969. Kept completely under wraps until the Earls Court show, it was unveiled on its own revolving plinth. The Amethyst was a complete departure from normal tourer design: its almost square profile was dominated by the use of vacuumed shaped plastic panels for the front and rear ends. The Amethyst's exterior had been designed by the David Ogle organisation, which was responsible for the stylistic lines of the Reliant Scimitar GTE sports car. Tom Karren, the actual designer, was responsible for the Conran Group's Siddall Delta some years earlier. The Delta's profile was very similar to the Amethyst's but the interior was a big let down and it was dropped pretty quickly. The Ogle group was responsible solely for the Amethyst's exterior, while the interior was left up to the Eccles designers. This pooling of ideas gained public acceptance. The Amethyst construction method usually had the chassis ending just at the axle. The 14ft shell was supported by wooden floor joists and the roof.

The big Sprite Countryman, also from 1965. I say big, because, in those days, at 17ft, it was considered a large tourer, more suited for static use.

Monocoque construction saw moulded plywood box sections along the roof edges and at floor level. The floor was dropped between chassis longitudinals, allowing extra headroom without increasing the van's height. Eccles designers found this was stronger than conventional chassis design, but after twelve months went back to the full chassis design. The 1971 Eccles went over entirely to the Ogle design, apart from the 16ft Sapphire, which retained the 1963 look. The Sapphire, though, would follow the Ogle design by 1972, and the Eccles uniformed profile was firmly established. The odd thing about the Ogle-designed Eccles was CI's claim that its design was futuristic as, by 1975, the Eccles profile was looking decidedly dated. Its boxy appearance gave way to a rounder and more conventional look.

Following in the Amethyst's shadow in October 1969 was the Fairholme 425. It, too, was based on a 14ft shell, and had integrated gas storage just near the axle line (an idea adopted by ABI in 1986). The 425 was a four berth tourer based on the exact layout of the Amethyst, but both exterior and interior were toned down to suit the Fairholme buyer. Even though the plastic mouldings used were not dissimilar to those of the Amethyst, the 425 was distinctive in its own right. Aimed at the upper clubman range, the 425 came with a shower, plus tip-up basin and vanity unit. The 425 also had the Electrolux RM10 fridge fitted as standard. Unlike the Amethyst, it only ran for two seasons before fading into oblivion. Fairholme had now found its home firmly at Newmarket, and with the New Dawn range established properly since 67, Fairholme was enjoying particular success in clubman circles. With the exception of the 425, little change was made for 1970. In 1971 the CI design department changed interiors and exteriors. The Fairholme design had become staid, with sales falling in 1971. For 1972, CI did some serious work on the Fairholme range and came up with the revised tourer line-up. Dropping New Dawn as a model designation, CI chose bird names such as Chaffinch and Goldfinch,

The Eccles Avenger from the Sprite works. This 1961 version looks dated, but Reg Dean would change all that the following year.

Reg Dean's first successful touring caravan design, the Eccles from 1962, gave the brand a much needed shot in the arm. Clean, modern exterior lines continue internally too.

Interior shot of the 15's front dinette. Roof locker storage wasn't a strong point.

last used by the old Fairholme company in 1961.

In an effort to grab a bigger share of the clubman market, CI decided to install hand polished, dark wood furniture in the interiors, finished off with satin-brass handles. Fairholme's new luxury furniture was manufactured in the new furniture section of CI. Oak veneer wall lining and tweed-effect upholstery, plus a heavy-duty carpet, gave the Fairholme interior a luxury feel. Fairholmes were now firmly established as the company's top of the range tourers. Quality, though, would have to improve within a few years as the Fairholme brand suffered build problems, but more about that in a later chapter. As Alper had previously done with Eccles and Fairholme, Bluebird tourer production was eventually brought to Newmarket in 1969. This move resulted in the Bluebird name being dropped from the tourers, while the model designation name Europe was adopted as a brand. Bluebird was now only manufacturing holiday homes back at Poole, and the Bluebird name confined to static units only. The CI Europes had changed very little over the years, and followed the Sprite's profile, although not to the extent of the Sprite price bracket. In an effort to distance the two brands, although they were made almost alongside each other, the CI design team realised that drastic marketing and design differences were needed. So, although production of the Europe models settled in at Newmarket, the 1970 range was to be the last.

For the 1971 season, the CI design department took the old Europe range by the scruff of the neck and dropped the name Europe (re-christening it the Europa). The design had new, box-type continental exterior and interior looks. The Europa took advantage of the CI plastics division, using the material extensively for the roof and end panels. Tom Karren was again involved with the CI design team (the Europa being his concept) and production involved plastic moulding techniques which were very complicated for

Fairholmes were still being made in Cardiff when this shot of this smart-looking New Dawn 15 was taken.

All furniture in 1969 Sprite caravans (except the Colt*) is made from afrormosia. Result—an overall appearance of appealing elegance and warmth. It also blends perfectly, with the curtain and upholstery colour schemes.

All work surfaces have a melamine finish to resist stains and heat. Just an occasional wipe with a damp cloth keeps them sparkling clean.

Felt-backed Vinyl flooring is one of the new features incorporated into Sprite 1969 design. Result—extra warmth and wear and an attractive appearance.

The Sprite kitchen unit gives plenty of space, drawers with improved handles—the top one with a plastic formed cutlery tray—and Sprite patented simmer taps, too.

Dense quality foam is the basis of Sprite upholstery. Unruckable, the seats easily convert into relaxing, sleep-inducing beds at night.

Unobtrusive locker handles, kitchen drawers that don't spring open during journeys, lightweight construction tables—these are examples of Sprite's attention to interior detail.

• The Colt is the basic Sprite model, and does not include such features as: hydraulic hitch, awning rail, panel jointing or wheel arch pressing as shown.

Sprite features are shown in this 1969 sales leaflet on the famous Musketeer model. Basic features by today's standards.

that time.

The interior of the new Europa was the work of CI designer Alan Boyce. The modern furniture blended in well with the soft furnishings and bright colours. Boyce gave the washroom a vanity unit and, for the van's living area, a large roof light and vent, an idea possibly copied from Dean's Lynton range. In fact, the interior was not unlike the Lyntons of the time in a number of ways, and its clean, modern looks were a breakaway from the norm. To add a small continental touch, the Europa came with an eye-level grill. However, this didn't win UK buyers' approval and was dropped the following year.

As the new Europas settled into their first year, dealer feedback suggested that the Europa design was perhaps a little too "boxy" and continental for UK buyers. Worried by the fact that sales had been "disappointing," CI set out to redesign the Europa, while still holding on to the original concept. For 1972, CI came up with a new, softer profile achieved by replacing the plastic panels with aluminium ones and reducing overall height.

Major changes included fitting a tinted Oroglass acrylic front window, while other exterior alterations were the use of two colours and integral pressed-out body louvres. Interiors were based on the 1971 design, but storage space was given greater emphasis by adding large shelves over pelmets. As mentioned earlier, the eye-level grill was dropped in favour of the conventional design. The 9ft Eurocamper was designed with the BMC Mini and the Hillman Imp in mind, and was set to continue with principally the same profile and market sector. The cheap and cheerful lightweight Eurocamper lasted until the 1974 model year.

Some years earlier, in 1967, the Sprite range was augmented by the Colt. Like the Eurocamper, the Colt was a basic tourer, even to the point where it had only two corner steadies and sealed, car-type, non-opening rubber window frames at the sides. CI further developed the Sprite range, and by 1969 had introduced the Alpine two berth. Manufactured with the door on the continental side, the two berth Alpine was basically an export model made available for UK buyers. It carried on into

1973 saw a new sleeker Sprite, the Alpine C, which cost £505. It was to be the last year that the Sprite brand had no direct competition.

the 1970 model year but didn't appeal to rally goers and was dropped from the range. Sprite then launched the Cadet, its smallest and lightest tourer produced to date.

The Cadet was aimed fairly and squarely at the small car owner, which was the popular Mini. The Cadet's 8ft 6in body length had the door located centrally at the rear. A small wardrobe, basic kitchen and compact front dinette saw this tourer sell at £227. Its novel large plastic roof vent let in plenty of natural light (earning the nickname "the bubble top") which also helped keep the price competitive. Accommodating two, or three at a pinch, it was intended for the first-timer and ex-camper alike. After one season, however, the Cadet was put on hold, and launched again for 1971 with a new sleeker shell.

CI's Sprite range had evolved from the early 1960s, the profile altering only slightly over nine years. Sprite not only launched the Cadet for the new decade, but also gave its Sprites a new look. Changing the exterior colour from dark green to almost peppermint was only part of the new Sprite treatment. A change of profile was implemented, beginning with the 400 and Alpine. The new "Sleekline" looks gave the Sprite a far more modern appearance, while maintaining a distinctive style. Interiors had new furniture, treated with a special heat resistant finish, along with new foam carpet.

After the first year, the new "Sleekline" design was transferred to the Major and Musketeer in 1971. In 1972, CI decided to improve the Sprite range even further. Smooth body panels were used and, for the first time in years, the famous green Sprite colours were replaced with a light tan centre band and white lower and upper panels. The Sprite name was boldly displayed on both front and rear panels. CI designed and used new lightweight furniture with rounded corners, made from heat formed hardboard. The bare white walls used in previous years were now discarded in favour of a slight pattern to give a warmer, less stark feel. Sprite popularity was assured - at least for the time being.

Making the Sprite tourer feel just a little less cheap and cheerful, but still offering value for money, would be something on which CI would concentrate over the coming years. Although CI was now a very successful company with many interests, it did experience a drop in profits, caused by production problems in 1970, on some of the new 1971 UK vans. Component problems, along with the troubled North American operations, contributed to the dive in profits. CI tried to

revamp its American models and brought in managing director John Moody from the company's South African operation. However, the American side of CI was still losing money, and was eventually abandoned. It is at this point that we close the CI story, although it could command a book to itself. Sam Alper's CI continued to make innovative tourers, but competition from the newly-formed Ace Belmont International (ABI) and other Hull companies such as Swift, along with over-production caused CI to lose its market share, and more money, by the late 70s.

The following is a concise history of the key events in the expansion of the CI dynasty and an account of some of the towing records it broke.

1963 CI founded.
1964 Sprite Caravans Pty (South Africa) joins CI Group.
1964 Wilk Caravans (Germany) joins CI Group.
1965 Africaravans (Gypsey, South Africa) join CI Group/Fairholme Caravans (UK) joins CI Group.
1966 Group HQ moves to Saffron Walden/ Queen's Award to Industry for exports from UK/Burdens (Accessories) joins CI Group.
1967 Harrison Bros. (Steels & Plastics) joins CI Group.
1969 CI Sport tourers launched (German Sprites).
1971 Tour–A–Way Caravans (South Africa) joins CI Group/OBI Awnings (Denmark) joins CI Group.
1972 Munro Caravans (New Zealand) (Oxford & Crusader tourers plus Sprite) join CI Group.
1974 CI Safari (Germany) launched.
1975 Production started in Italy (Riviera)/CI Finance formed/Albatross Industries (Australia) joins CI Group.
1976 Production started in Sweden (Savsjo)
1977 Coventry Steel Caravans joins CI Group (UK).
1978 Fairholme Tourers move production to Bury St Edmunds.
1979 South African subsidiary becomes public company as CI Industries Ltd/ Cooper Coachworks (UK) joins CI

The "Mini breakthrough" baby Eurocamper ran for almost 20 years from 1965. It weighed in at 8cwt and cost an incredible £253.

The Bluebird Europe four from 1969 was the biggest in the range at 16ft. The following year the Europe was moved from Poole over to Newmarket. Note that the lines are not dissimilar to those of the Sprites.

Group/Flambo (Pty) Limited joins CI Industries/Premier Cargo Vans (Pty) Limited joins CI Flambo/CI Autohomes GmbH start German production.

World speed records held by CI tourer brands

Caravans International broke more records and endurance runs with its touring caravans than any other manufacturer to date. These are just some of the numerous accounts and dates of records achieved by Sprite tourers from 1960 up to the early 70s when production was finally halted.

1960 Outfit: Sprite 400/Mini complete 825 miles from London, Cardiff, Edinburgh, Belfast and Dublin in 38 hours 54 minutes.

1961 First British caravan to be towed through into Moscow was a Sprite Musketeer. Outfit: Sprite 400 and Mini covered 1083 miles in 24 hours around the Snetterton race circuit averaging 45mph.

1962 Outfit: Sprite 400/Morris 1100 completed 1114 miles in 24 hours with an average speed of 46mph at the Zandvoort Grand Prix circuit in Holland.

1963 Outfit: Simca 1000 and Sprite 400. Was driven over the same route as the Alpine Rally. In five days 62 Alpine passes from 1800ft to 7000ft above sea level were covered, 1400 miles at an average speed of 25mph.

1964 Outfit: Sprite 400/Landrover. This outfit was taken over the Sani Pass in Africa and proved the build quality of Sprite tourers. A Sprite Alpine and Ford Zephyr 4 completed seven out of ten mountain stages of the Monte Carlo Rally at an average speed of 27.3mph. Still in the same year, a Jensen C-V8 towed a 400, Alpine and Musketeer at over 100mph at Duxford aerodrome. The Musketeer set the world caravan towing speed record at just over 102mph! On the same day an Eccles Moonstone was towed at 97mph by the same car.

By 1973 the Fairholme range was well and truly targeted at the clubman. Check out those altogether smoother lines. The whole picture comes from a Fairholme brochure: the picture frame, presumably, indicating a masterpiece.

1966 After a year's rest, CI was at it again! Staged in Italy at the famous banked Monza circuit, a 16ft Sprite Major towed by a Ford Zodiac covered 1689 miles at an average speed of 70mph in 24 hours. In an effort to show how easy it was to tow a Sprite tourer, two drivers with no towing experience were sent on a tour round Britain. Using a Sprite 400 and Hillman Imp, the pair covered 2068 miles in just over 69 hours.

1967 Outfit: Sprite 400 and Hillman Imp. This outfit did a 1804 mile tour of Portugal at an average speed of 35mph and in a time of 51.43 hours. Using pedal power, two girls rode specially adapted cycles to tow a Sprite 400 around London, this bizarre exercise proved just how well Sprites could be towed. Porsche of Germany towed a Sprite Musketeer at speeds of 88mph, and even managed speeds of up to 74mph up 1 in 8 gradients.

1968 Outfit: Prototype Eccles Opal and Triumph GT6. Towed over 2000 miles on European roads, managed to keep up an average speed of 47.6mph, completing the run in 42 hours. As chief competitor in the British Caravan Road Rally, CI swept the board with six wins. No other tourer manufacturer ever beat this winning performance. A prototype Sprite Musketeer was towed over 100,000 miles in South Africa. Towing down dirt roads between Ghost Mountain, Mhuzi, in Zululand, to

Interior of the 1973 15ft Goldfinch, one of Fairholme's most popular models. Natural wood veneers with brass handles give luxury clubman appeal.

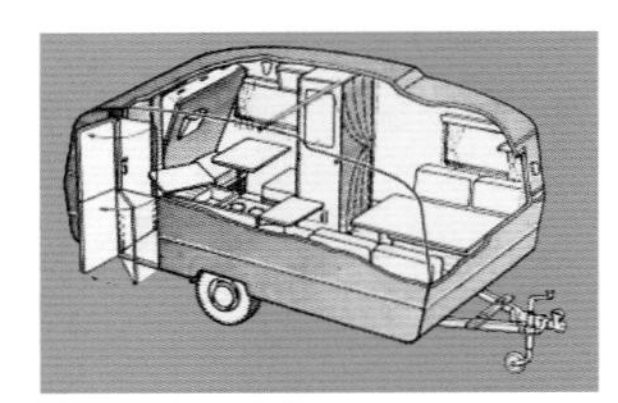

Manzini in Swaziland and back, the outfit could average speeds of 50 - 60mph. The Sprite came through it with little damage.

1969 Outfit: Alfa Romeo 1750 and Eccles Opal. Liz Firmin towed the Opal from London to Rome in 24 hours at an average speed of 50mph, covering a distance of 1188 miles. After a short time away, Sprite went back to the Monza race track. Here, it took a 400 1970 prototype and towed it with a Ford Cortina 1600E for 5000 miles in just short of 83 hours. The Alpine, again another 1970 prototype, was towed by a Ford Corsair 2000E over the same distance, doing it in 81 hours.

1970 Outfit: Europa 390 and Austin Maxi. In the hands of lady rally driver Rosemary Smith completed a round trip of Europe.

1971 Unofficial Sprite entered the 1971 Baja 500 in California, reaching speeds of 85mph on dirt roads!

1972 Outfit: Eccles Topaz and Chrysler 180. Driver Graham Birrell drove the Monte Carlo rally route at a speed of 54.6mph. CI took a Peugeot 405 and Europa 390 on the East African Safari route, covering 1000 miles.

1973 In Sweden a Mr Bo Frisk claimed a world record for ski-caravans at nearly 33mph, towing a Sprite Alpine. CI and Sprite founder, and then Chairman, Sam Alper, drove a Mini and Sprite 400 outfit over 10,000 feet above sea level, in the Sierra Nevada.

This seemed to signal the end of the CI endurance era. As you read through the rest of the chapters, CI is strongly challenged by competition and poor sales and heavy management, until finally, the Alper empire crashed.

When Tom Karren designed the Amethyst it caused a sensation with its up to the minute good looks. CI was pleased. By 1973, though the design had been tweaked here and there, it remained basically the same.

All-new Amethyst

Two years on the design had become dated and sales were at a low ebb. CI redesigned it for 1976 with this far more pleasing profile, which looked very Elddis. This is the Topaz 12 two berth.

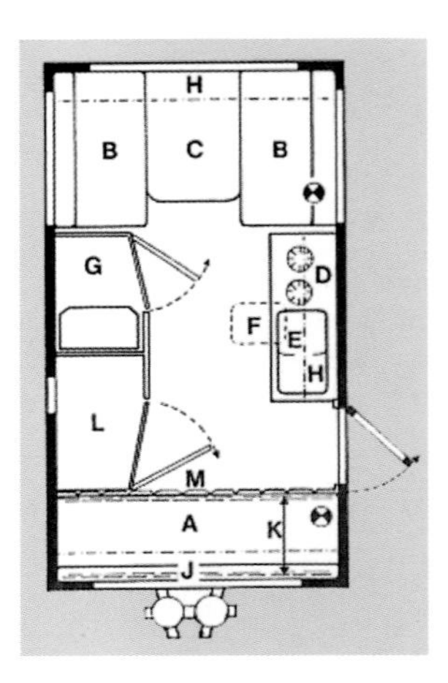

This year the Simca 1100 and CI Eccles Opal caravan proved unbeatable in the Team Section of the Esso Challenge Cup.

They also drove off with the Parking Award.

And came second overall in the Rally.

Results like these prove the Eccles Opal is a great caravan. And Simca is a great towing car.

Simca/CI are making a habit of winning caravan rally awards—last year outright winners of the Caravan Rally.

Send now for our free booklet on the Simca or Eccles range . . . they're all great towing combinations.

The British Caravan Road Rally was a popular event. CI swept the board in 1968, and had a good 1969 event as well.

The industry comes of age

Just as the late 40s and early 50s spawned new touring caravan manufacturers, and witnessed the demise of some older makes, the UK industry in the 60s experienced a huge expansion in retail as well as the number of suppliers and manufacturers. We will look at how tourer dealerships expanded and sites became far better equipped and organised in their approach to the caravanner.

Prominent names of the previous decade, such as Paladin, Berkeley, Raven, Winchester and Bampton, to name but a few, eventually disappeared from the manufacturing scene. The clubman market, too, underwent major changes. We will cover clubman luxury manufacturers in the next chapter, which is devoted to that almost magical era of hand-built craftsmanship.

In the north caravan manufacturing mushroomed, especially in the Hull area where the industry was founded by Willerby, followed some years later by Lissett, Cresta and Astral. Scotland continued to be dominated by Thomson, a company that had grown from virtual unknown to well-respected manufacturer of tourers. Not only was it Scotland's only volume caravan manufacturer, for a short time in 1961 it also built 150 motorhomes, making Thomson Scotland's only motorhome manufacturer, too. Bailey, the west country manufacturer, had stayed in the background since being founded, but, as time went by, the company steadily built up a reputation for quality.

Just as the manufacturing side of the industry was growing, dealers who had established themselves early on were expanding, opening up adjoining dealerships. Dealers like Harringtons in Delamere, well established since 1924, found the next ten years to be a period of expansion, adding two more sales grounds by the early 60s. Burtree Caravans in the northeast had developed another three sales grounds (Burtree became Barrons Great Outdoors in 1994).

In the south, Jenkinsons (no connection to the author) had display grounds in London and Taplow. The Gailey Group spread its wings by going nationwide, with a Gailey dealership in every part of the UK. In fact, the retail trade was expanding at a greater pace than the manufacturing side, with garages also taking on caravan sales as a sideline. Even manufacturers, in some cases, either stopped production or did both by becoming retailers too. Burlingham, Safari, Mount, Pilot and even Thomson moved into this growth area. Caravan dealerships would spring up almost anywhere a catchment area existed.

The Caravan Club was also benefiting from the increasing popularity of the touring caravan. With members joining faster than the Club had ever experienced, the start of the 1960s saw the Club begin to change its structure. One major change was the formation of the Club as a company, which was to be known as the Caravan Club Ltd, dropping the "of Gt. Britain and Northern Ireland." By 1963 the club had

Astral's factory had grown as demand for the company's caravans soared. The Hull maker was just about the largest in the area with its Clough Road production plant churning out caravans for home and export.

launched its own official newsletter: "*The En Route Magazine*," a title that would last 35 years. At the end of the 1960s, club membership totalled just over 100,000. By the 1990s this figure was nearer 300,000, with the Camping and Caravanning Club not far behind.

With the industry's expansion, several publications dedicated to both the trade and caravanners sprang up. *Modern Caravan*, derived from Berkeley caravans' news magazine in the 50s, was second in readership figures to the long established *The Caravan* magazine. The *Trailer and Trades Journal* was a trade paper which was later incorporated into *The Caravan*. *Caravan Life*, a northern publication, was taken over by the Haymarket Group. By 1967 *Caravan Life* was dropped (relaunched 20 years later in 1987), with *Practical Caravan* taking its place and becoming the UK's best selling caravan publication.

Getting back to caravan manufacturers and launching of models for the new decade, in September 1959 the first international caravan show was held at Earls Court. Visitors from overseas came and placed orders with the UK manufacturers. Just for the record, the first caravan sold at the show was a Willerby Villamobile, an end-bathroomed mobile home sold by the Pathfinder company.

Over 87,000 visitors came to view the 161 caravans (manned by 32 dealers and 55 other connected caravan exhibitors) for the new 1960 season. This exclusive new show gave the industry kudos and began to dispel the cottage industry image. Just five years later, this event had become the caravan industry's showcase, with over 113,000 people attending (a different story to the Earls Court Show of the late 90s, which struggles to get 40,000 through the turnstiles).

Touring caravan manufacturers were enjoying a boom era, and none more so than those in the East Yorkshire area. Most of this chapter, therefore, will be devoted to these new concerns because of their importance in making the industry what it is today.

The main attraction of the east Yorkshire area was the docks, which facilitated the import of materials and the all important export trade. In some instances the latter meant the docks were littered with caravans of all shapes and sizes. What a sight that must have been!

Astral was Hull's newest manufacturer. It had only been in business since 1959, but had already set production targets of 2000 units for 1960. Owned by the Spooner family, Astral was part of the large timber and civil engineering company (Spooner Group), and was keen to establish itself as a major player in caravan manufacture. Early models were based on living/holiday units, with tourers initially playing a small part. It wasn't long, though, before Astral was finding its feet in tourer manufacture too. Most designs were greatly influenced by the Spooner family, who often used their caravans, and appeared in Astral's sales brochures.

The first serious tourers came with the Travelite, a 11ft 6in, four berth tourer, which was joined by the Cameo in 1961. The Cameo featured a new lightweight chassis, but, even so, the 14ft body length commanded just over 14cwt unladen on the scales. The Cameo was dropped for 1962, making a brief reappearance in 1963. The Cameo name came into its own later on.

The Leda, a 17ft tourer, was almost American in style, with a bold exterior featuring steeply sloping end walls and two entrance doors. It was a brave attempt, but a failed one. Astral followed it with the Sunrise in 1964. Rear doored, this 14ft, six berth model featured a Luton top housing a double bed. The Sunrise also had the old Berkeley Messenger design of a fold-out rear panel which formed an extending roof. The Sunrise was also very influenced by American designs but, like the Leda, never caught on and was dropped at the end of 1965. After a state of almost experimental designs and models, the first of the popular Ranger line-up of tourers was launched in 1964, establishing the Astral tourers into the 1970s.

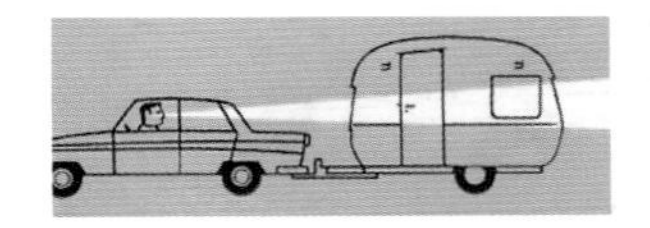

Astral's Sunrise model came complete with a Luton top. Influenced by American design, the Sunrise lasted only two seasons. Before it went on sale, the Spooner family had toured the continent in one.

Willerby, who was also very big in holiday home caravans and living units (just like newcomer Astral), was trying to identify itself with a tourer range that would take it into the 1960s, even though the Robin off-shoot of the company produced a small range in nearby Beverley. Willerby was very big in exports, and this, as was the case for many manufacturers of the time, played a vital role in the overall design of its product.

In 1959 Willerby launched a micro tourer, the Tip-Top, designed for small car towing. The 9ft 6in Willerby was a three berth, boasting full road lights. For 1960, it became an export only model, available by special order in the UK at a cost of £225 (or, if you preferred, £170 for a DIY shell only).

The Willerby Kite was a 12ft, four berth, two door design, incorporating a centre dinette, end kitchen (still a popular feature) corner loo compartment and a single width bed at the front. The Kite didn't need any major modifications for export, which played a vital role in keeping costs down and the Willerby competitive overseas. Willerby described the interior, as being "up to date with gay curtains and upholstery."

In 1964 Willerby launched the successful Dandy tourer. Its sure-fire export potential, with the door on the offside, saw the cheap, curvy, fully insulated, lightweight 10 footer take off on the home market (to such an extent that the company added 12ft and 14ft designs to the range). The fact that exports influenced the Dandy didn't in any way detract from sales potential at home.

Cresta, which had become another founder of the Hull caravan manufacturing centre in the 50s, saw its tourer range being reduced. The once popular Trans-Continentals were stopped and replaced by new models. The popular little Kozy 10ft model was eventually dropped in 1961. The reason for this was that, just as Sprite had begun to expand by acquiring Eccles, the Rigby brothers over in Wigan had seen their Pemberton range of tourers and holiday homes take an increasing slice of the market. They approached William Turner, Cresta's managing director, with a view to taking over the company and its Clough Road premises in Hull. Turner, who also owned Turner's Trailers in Rochdale, the caravan distributors, bought his fellow shareholders out and made the decision to join Pemberton Caravans. With this, the Cresta range could infiltrate new export markets, and with increased capital, could also increase Cresta production.

The deal was made in September 1961. Pemberton dropped its tourers, concentrating instead on the holiday home market. Cresta's last tourer was the 14ft Scout. A scaled-down static holiday home, it was launched for the 1963 season with a price tag of £330. Its triple deep front windows gave good see-through vision (a big selling point in tourers up to the early 70s) but it wasn't enough to sell the rather ugly looking tourer. By 1967, the Cresta name had been dropped for good and the units were marketed as Pemberton, along with the

Blackpool-based Dovedale.

Although Willerby and Astral now had a firm foothold in caravan manufacture in the East Riding, it became obvious that the ever-expanding market for tourers and holiday homes could not be filled by just these two companies. It is here that we now see new, smaller manufacturers being spawned from the big two. Firstly, the beginnings of the ABI empire. Terry Reed, who had been with Astral as a joiner, left the company in 1962 and, with a capital of £2000, set up Ace Caravans at Colonial Street, Hull (no connection to the pre-war Ace Caravans). Reed produced just two models for his first season, one of which was aimed directly at export markets. The Ace Ten cost £275, while its sister, the 12ft Continental, cost £350. Although not the most exciting tourers to look at, they did represent value for money and were quality-built units. Reed found an importer in Holland who, eager to take on his Ace range, bought Reed's first batch. Clearly on to a winner, he started building five tourers a week. In fact, Reed's Aces could only be found on Dutch dealers' sales grounds as all his production was allocated for export.

Following increased production, and showing for the first time at Earls Court in September 1963, Ace's four-tourer range attracted much attention. The vans were equipped with a hand-operated water pump for the kitchen, as well as fitted carpets, 12 volt electrics, gas lights, touch-up paint and crockery, and a smarter design outside and in.

Innovations included the use of a curved toughened glass front window and an unpainted front stucco aluminium lower panel. This was designed to absorb stone chips. Needless to say, this idea was quickly adopted by other manufacturers, as was the front window design by Robin, Knowsley, Colonial/Olympic and A-Line. By the 1966 model year, familiar names, such as Rallyman, Globetrotter, Pioneer, Ambassador Airstream and Viceroy, were launched and all became very successful models (some ABI 1999 models still use these names).

Reed couldn't keep up with his orders, so a move to larger premises at Oak Road in Hull was implemented. The caravan press raved over the mid-priced Ace tourers. Reed had, unknowingly, sown the seeds of what was to become a major caravan manufacturer. By the late 1960s Reed had moved his production again, to Swinemoor Lane, Beverley, this time, and expanded the factory. He began to get

Early Swift 10 model cost £289 in 1965. The Smiths began in a very small way. At the time, CI had already claimed to be the largest tourer manufacturer. However, in 30 years the positions were completely reversed (Courtesy of The Swift Group)

involved in the holiday home caravan market for the 1969 season, and also marketed a five-tourer export range named the Star, making it available for UK customers. The Star lasted only one season, though.

The Yorkshire-based Ellbee window and door company supplied all the Hull caravan manufacturers and Ace was no exception. For a short time in 1967 it became the fashion to fit louvred windows in most tourers' entrance doors. This soon fell out of favour though due to draught and security problems. Pre-finished aluminium panels became widely used at this time, which meant that factories didn't need spray shops (a fire hazard for many caravan manufacturers). Ace was one of the first to use these off-the-peg, pre-painted aluminium sheets, which became a clear winner over hardboard by the late 50s and early 60s.

Swift, which had started life in 1964 (two years after Ace) as a one-man band operation in a small garage at Hedon Road, Hull, was soon joined by Ken Smith. Smith saw the company's full potential, since tourers were in greater demand than ever before. At first, only four tourers were available, built by a small team of eight men who were also involved in sales, material purchase and delivering the finished tourers on their weekends off!

From its beginnings in that small shed, the company developed, took on more buildings and increased production capabilities and workforce. Ken Smith and his wife Joan were now very busy making Swifts which, for 1965, came with independent suspension, full insulation, buttoned foam mattresses, hand -

operated water pumps and fitted carpets.

Swift, like Ace, evolved with an easily recognisable profile: triple deep front and rear windows, sapele finished furniture, and an under-floor larder. Exports were also successful for Swift, especially in West Germany, mainly due to the new 1966 models, which sported long-running model names such as Alouette, Silhouette and Baronette.

One notable model in the Swift range, which was brought out for 1968, was the Rapide, a tiny 8ft 6in tourer designed for the smallest cars (launched again in 1987 as a 12ft budget tourer). Swift produced mid-priced tourers, which meant that a fridge and a full cooker were optional extras, adding a grand total of £54 to the selling price!

The next manufacturer to emerge from the Hull nucleus was Mardon Trailers, founded just after Ace in 1962. Les Marshall, who had worked with Willerby buying materials, had several ideas of his own about how tourer and holiday home caravans should look and perform. Marshall founded his fledgling caravan company with the Mardon Mini. Measuring just 9ft in length, its compact size could accommodate four, albeit at a squeeze.

This area of the tourer market had seemingly been spurred on by the launch of the Mini car. Other manufacturers that devoted models to this market sector included Colonial, Dovedale, Cygnet Waterbird, Fisher and Penguin. Export demands were so strong that early Mardons were finding ready-made markets. The 10ft Minor and Major models were aimed very much at the continental buyer, and didn't look very different to the Willerby Dandy.

Although Marshall had borrowed more than the odd idea from his old company, resulting initially in almost clone-like designs, Mardons would eventually find their own identity with the Classique range. Originally almost square in appearance, they would, by 1968, take on a more distinctive profile, sporting stainless steel front and rear lower panels topped off with a heavy patterned aluminium exterior. Interior design, though, would become the Mardon trademark, with soft padded vinyl walls and roof (known as insulux) and teak-effect finished furniture.

Mardon was well established (within quite a short time) when disaster struck the factory. In late 1968, a fire destroyed the soft furnishings department resulting in loss of production for a short while and putting a £300,000 export order in jeopardy. It may have been a disruption, but it didn't slow Mardon's growth.

Not too far from Mardon, at Pulman Street, Spring Bank West, Hull, another new manufacturer was being formed. The Robinson family, who had built a tourer for their own use, saw the caravan manufacturing industry booming, especially in Hull, and took the plunge to enter what was fast becoming a localised industry.

Mr Len Robinson began the Silverline Caravan Company, which launched its first caravan in April 1965. A staff of just four produced just one caravan a week. Three years later, 40 staff were on the payroll and turnover was £250,000, helped again, no doubt, by exports (most early Silverlines went to West Germany).

The 1966 Silverline range boasted two tourers plus a couple of holiday homes. The Pullman Royal, a 13ft four berth, looked remarkably like the up-market Bessacarrs of the time. Len Robinson introduced the Pegasus for the following season, a 12ft two/five berth tourer costing £350, under-cutting Swift and Ace who had similar models. The Silverline range came with fitted carpets and a hand-operated water pump, along with the then quite basic standard equipment.

Silverline, though, became more involved with holiday homes and kept the tourer range to a minimum, although it did produce some large "just towable" caravans, one being a twin axle unit. The Silverline Pullman Princess found its niche in the showman and travelling circus market; the company even supplied a single axle model to Penny Smart, granddaughter of circus king Billy Smart.

Still new East Yorkshire manufacturers kept popping up like mushrooms. Alpine Coachcraft, a small company based near Ottringham, just outside Hull, and apparently using an old BBC station as its works, began in 1965 by producing the Super model range. Alpine initially launched two mid-priced models but, for 1966, nine tourers were available, all in various layouts and sizes ranging from 10 to 14ft (again, these layouts were export orientated). Within a short space of time, the company changed its name and logo to A-Line Caravans. A-Line soon became a hit with the buying public, its tourers featuring the Ace idea of a curved front window, and the then ever-popular boat roof design (so called because it resembled an up-turned boat). A-Line added blue metallic-effect lower and exterior wall

Bristol-based Bailey had quietly manufactured tourers since 1948. The company had built sales slowly in the UK and abroad. The 10ft Maru proved a popular four berth van from its launch in 1964. This is a 1965 model.

THE maru 10′ 6″ LIGHTWEIGHT CONTINENTAL TOURER

PRICE £298 ex-works

The vans of Bristol-based Bailey had, by 1969, begun to become more popular. This 22ft Montana from 1969 boasted a loo compartment as well as five berths and full-blown end kitchen. Unladen weight was 30cwt.

panels, which made the vans quite distinctive. Problems arose in later years, though, when the paint sometimes flaked off after long-term weathering.

Interiors were conservative, with predominately yellow soft furnishings and the usual photo veneer. By 1968, the company was producing eight touring models and, like

From 1962, this Fisher Holivan Junior put touring caravan ownership within reach of most car owners. At 8ft x 5ft, compact tourers such as this are what Fisher became famous for.

Ace produced some of the best-looking touring caravans on the market, such as this 16ft Viceroy from 1968 costing £499. Uncluttered looks, and bowed front window were Ace's distinguishable markings.

everybody else, went down the holiday home path, basing the design on the tourer range. The Super holiday home vans became very popular, with their smart lines and those metallic blue body panels.

A-Line soon expanded and became, for a short while at least, a formidable force in the industry in the 1970s. With caravan production in 1965 running annually in the UK at around 50,000 new units produced and sold, even the unlikeliest of companies seemed to see huge potential in touring and holiday home caravan manufacture.

One such company was the large Cosalt organisation based at Grimsby. The company had many operations under its giant umbrella, one of which was manufacturing insulated containers. It was this side of the business that was abandoned, due to the decline in demand, in favour of the new caravan company named the Humber Caravan Co. Two ex-Astral personnel, Bill Boasman (Astral's works manager) and Barry Holmes (Astral's sales office manager) joined Humber's managing director, Len Funnell, to set up the new division.

The new team designed and built several prototypes before finally going into production on May 25th 1966. A marquee was erected for use as a temporary factory and it was here that the first Humber Abbey was built. Named the Twelve, this four berth tourer cost the then princely sum of £354. Five tourers were coming off the line each week, produced by only ten employees. The first Humber Abbeys elicited interest from export markets virtually from day one, with the first 18 destined for Europe!

Marketed in the main to the mid-priced tourer buyer, the Abbey models didn't look too impressive, especially compared to the new Ace, Swift, Mardon and A-Line designs. Cosalt dropped the Humber name in 1969, due mainly to it being confused with the car manufacturer, and so the Abbey Caravan Company was formed. Along with the new name came a smart new image which gave Abbey caravans new appeal.

Abbey products were very similarly priced to the other Hull-produced ranges and from its factory at Convamore Road, Grimsby, the company produced holiday homes as well as tourers. Soon after Abbey was established,

After only two years, Abbey tourers had established themselves with the Abbey De-Luxe. This is the 12ft from 1968 which cost £387.

Ace interiors were pleasing to the eye. Viceroy had a 7ft width (unusual for the time) and offered a loo compartment with washbasin, another unusual feature.

Cosalt decided to further its interests in tourer manufacture by acquiring, in 1968, Safari Caravans, the luxury clubman maker down in Stroud. This move left no doubt that Cosalt meant business, especially when, six years later, it acquired yet another manufacturer.

In May 1968, Derek Upfield, who had worked at Ace, left to set up his own caravan manufacturing business. Naming his company Riviera Caravans, he set about producing a distinctive range of tourers. The design featured a thrust-forward body shell with a profile not too dissimilar to the last Sport tourers of 1967 built by Summerdale. Upfield's tourers received good press reaction. Light oak veneer furniture finish and a quality feel allowed the 12 and 14ft five berths to expand to other layouts. A quote from *Caravan Magazine* said that "Riviera would be one of the few companies to survive the already over-supplied tourer market."

Marfleet Joinery was another of these Hull opportunists. In May 1967, company founder and owner, Mr A. Taylor - who had previously been works manager at Astral Caravans (Astral seemed unofficially to be training up new bosses for caravan manufacturing concerns!) - set up his factory producing 11 and 14ft tourer ranges. Sovereign was chosen as the brand name. With a design which was almost box shaped, the new Sovereigns found their niche due to solid build quality and a competitive price tag. Taylor wanted to see 70 vans produced each week at the new £50,000 factory complex, covering six acres, that he was having built near Hull docks ready for the following July. With a new production plant, Sovereign launched the up-market versions of its tourers, the Tiara, Coronet, Diadem and Herald, in 1969. With a new, distinctive and shapely body shell, plus additional items such as 12 volt lighting, a gas point and a hand basin with its own water pump in the washroom, Sovereign became a popular range at home and abroad.

Yet another manufacturer that came about in the booming late 60s was the small and obscure manufacturer, Haltemprice. The factory, on George St, Cottingham, Hull, made the 12 and 13ft Henderson models. Pitched above the mid-market, they cost £500-£525, but, kept just below clubman prices, the company's models didn't prove very popular. Haltemprice also built special units and for a short time did better concentrating on this area.

Although based at Brandsburton near Drifield, one maker that produced a distinguished, although short-lived range of tourers was Sabre Coachcraft. Started by Les Garside and Brian Johnson in 1967, its two model range increased to six in 1969. A distinctive feature was the Sabre's overhanging boat roof design, giving the impression of it being too large for the van's body. Two years later, in October 1969, the Sabre company closed down.

It's now time to deal with the "stragglers" of the Yorkshire-based manufacturers, such as those founded towards the end of the period covered by the last book. As I mentioned earlier in this chapter, Robin caravans had become part of Willerby and were being built at Beverley. Although a well established company it wasn't until the mid 60s that the vans took on a more pleasing shape.

As with the majority of manufacturers, Robin aimed its range at the middle of the market. Quality and sound design meant the Robin gained popularity and its now smart exterior resembled that of the Ace range. By late 1968, Willerby dropped tourer production altogether and, since both had now become part of the Burndene group, it was decided that all Willerbys should be holiday home units whereas Robin would concentrate solely on building tourers for the company.

Lissett, which had remained more of a manufacturer of mobile homes and holiday caravans, did produce a tourer range in the late 50s. It wasn't until 1966, however, that the Alpha tourer was launched. It was soon dropped, though, and in 1967 the Cuban range

The Lissett Cuban from 1969 had a profile similar to that of the Cavalier. Lissett was another early East Yorkshire maker which later went on to make holiday homes.

was reborn. It was probably export orders that kept tourer production going at the old airbase factory, with the Cuban finding favour in Scandinavia. Its exterior profile wasn't unlike that of the Estuary Cavaliers (box-shaped with rounded corners).

It's clear that the caravan industry in the Hull area was now possibly equal to that of its famous fishing industry, both in terms of turnover and the number of people employed. These expanding companies were now ready to take on Caravans International, the Newmarket giant which had the touring caravan market pretty much sewn-up and was seemingly unassailable. British touring caravans were now in greater demand than ever, and the market seemed almost limitless.

Moving away from the Hull area, it is now time to take a look at what was happening in the caravan industry in other parts of the UK during this decade of buoyant growth.

Valley caravans, manufactured by the Northern Pattern Company in the north east since the mid 1950s, aimed their tourers at just a shade above the middle market sector. By July 1964, though, the company ceased manufacturing caravans. Waiting in the wings, however, was a new company - one which would become one of the biggest success stories in the industry, next to Swift, CI and ABI.

Siddle Cook, owner of a north eastern haulage company, had been a keen caravanner since the 1940s and, in 1963 after owning several makes, he decided to design and build his own tourer. His keen interest in rallying meant he attended the Caravan Club's Longleat National Rally. In those days the amateur-built tourer (official name non-prop) was still a popular pastime amongst those who had the skills. The Club used to have a special competition for the best improved or built design.

The Cooks won first prize in the event with Siddle's luxury tourer, which had a built-in radio, record player and integral front spare wheel storage compartment. Cook also fitted his own auto reverse system, which worked off compressed air generated by the car. Standard auto reversing systems on tourers were still 11 years off!

Two years later, Siddle and his son Raymond started manufacturing tourers. Finding a name for the new company was difficult, though, until Ray Cook spelt Siddle backwards – Elddis Caravans was born! They set up a small factory at Esp Green, near Consett, Co Durham, where initially only a few tourers were built - the design wasn't inspiring and looked rather bland. This would change in 1966 when two, now very familiar, models were launched: Whirlwind and Tornado. The redesigned profile and interior quickly caught

North eastern manufacturer Elddis began in late 1965. By 1966 it was producing two tourers. The 10ft here, has the early profile of the later models.

The Elddis 12 has a hint of the clubman car cruisers of that era. Elddis also dabbled with light sailing boat manufacture. In those days the company would also build a caravan to customer requirements.

on and, with a value for money price tag of £395 for the Tornado, Elddis was quickly establised.

The father and son's work force were producing seven units each week and these were being sold as quickly as they were built. In March 1967, however, the Elddis success story was threatened in the form of a fire at the factory. Several tourers were lost but production was back on the go shortly afterwards. Expansion in 1968 saw the company build its Delves Lane factory, still the company's site today.

With demand high, Elddis launched the 14ft Cyclone and increased production to 45 a week. Smart distinctive exteriors, and light oak veneer furniture with rounded corners, became Elddis trademarks. Competitive on price while maintaining quality, Elddis carved itself a niche in the new tourer market.

Further north, in the Scottish borders, Thomson Caravans (its history is fully traced in *British Trailer Caravans & Their Manufacturers 1919-59*, also published by Veloce) had seen sales rise at a very steady rate, at home and abroad. By the 1960s, the familiar T-Line profile had evolved, along with the easily identifiable "Glen" models.

Many caravan manufacturers regarded Thomson as something of a "backwater" manufacturer, being based away from the mainstream manufacturing areas. However, this rather unfair, but somewhat inevitable image didn't stop Thomson taking its fair share of the UK touring caravan market. Nor did it stop it ranking sixth in the league of manufacturers. These figures, in 1963, could not have been imagined by Thomson a few years earlier. With seven tourers in the Thomson line-up, the Glenelg, introduced in 1963 in its 13ft four

The Fairview Laagan from 1964 poses for the camera with its smart white-wall tyred Hillman Minx. Success with tourers wasn't brilliant, with mobile and holiday homes being the main production at Harwich.

Lytham-based Colonial clashed head-on with Fisher with this almost clone-looking Galaxy model. Colonial even reckoned motorcycle towing was effortless and safe.

berth guise, gave the company a best seller. In fact, the 1964 record for production (2000 tourers) was broken with the Glenelg (by 1966 Thomson had broken it again with 3475!)

From the big 18ft family Gleneagle down to the Mini-Glen (launched in 1967 primarily for the Mini car as it was only 8ft 6in in length) Thomson had the market well and truly covered. The Glen 2/4, Glendale 2, Glennevis 4/5 and the Glenelg four/five berth were Thomson's most popular models. At one time every other tourer was a Thomson T-Line. Soundly constructed, complete with proper tongue and grooved floors, thick gauge aluminium and mineral wool insulation, Thomson was now leader in the middle price category. By 1964, the owners' club was formed with 700 members, making it the second largest in the UK.

Becoming a public company in 1967, Thomson was following in the footsteps of Sam Alper's Sprites. The T-Line Group had timber importing companies, a soft furnishings factory and a new holiday home factory. In 1969, the company opened its Canadian distributor (Thomson-York) which took over 104 units in the first six months. Abbey, Bailey, Cygnet (who manufactured there) as well as Thomson had all seen this as a new market area. However, this operation, for all four UK makers, later proved uneconomical and Canadian exports stopped. Scandinavia, Norway, Sweden, Denmark, Holland, France, Switzerland, Germany and Finland were export markets for the T-Line Group. The colder climates required Thomsons with insulated floors, double glazing and gas heaters as standard. Thomson was a very large concern and profits had risen from £53,694 in 1960 to £350,437 in 1969. We could devote the whole of this book to Hull manufacturers, Thomson and CI, but must now move on to other makers who deserve more than just a passing mention.

Knowsley, based at Appley Bridge, Wigan, Lancs, produced some fine tourers. Edmund Taylor, a poultry industry equipment manufacturer, produced Conway Caravans (and Conway Trailer Tents some years later) as a sideline. Taylor bought the Knowsley name in 1962 and dropped Conway to concentrate on the Knowsley range. By 1964, all the models had Greek influenced names with 1966

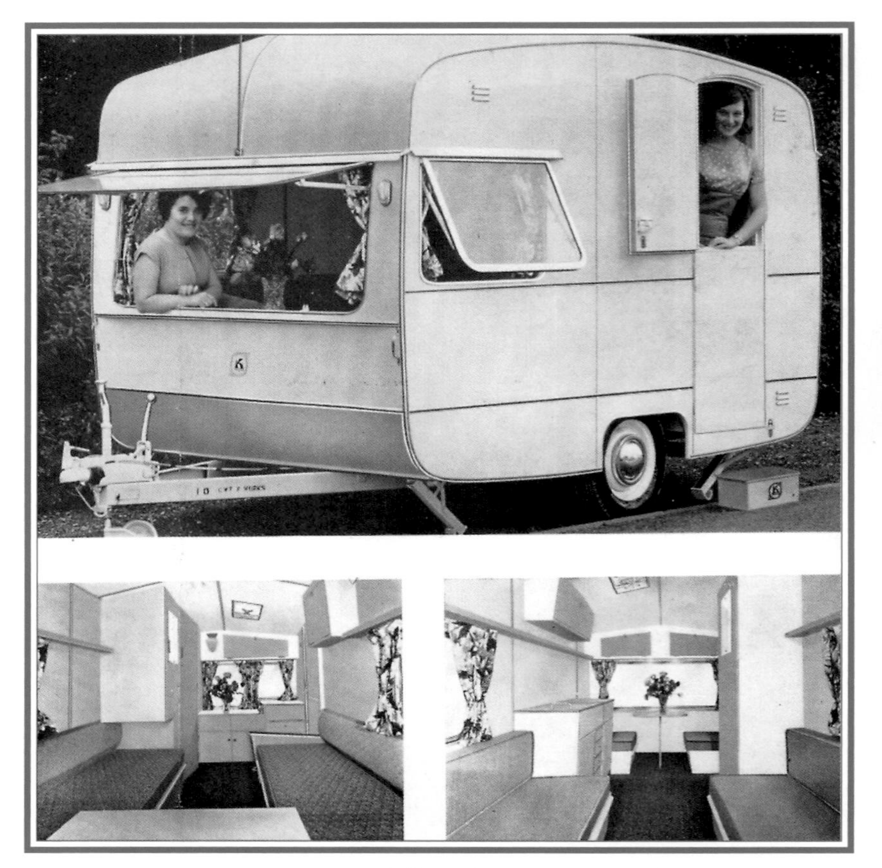

Knowsley had established itself by the mid-60s. The Eros became a best-seller in both two and four berths.

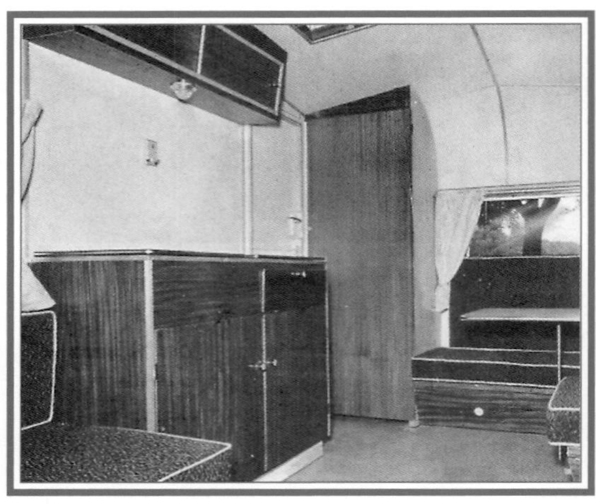

Export influenced, the Swift's interior had Afromosia furniture and came with a foot-operated water pump in 1966 - all for £357.

establishing the Eros, Juno and Hermes as Knowsley models, all of which were clubman-style tourers with spacious interiors and real oak veneers. A novel feature was the end kitchen layout, which meant that the kitchen unit could be adjusted (when on tow it was lowered for see-through vision, and put up when on site). This idea was also used on Swift and Welton tourers, though no-one knows who thought of it first. Knowsley was like Astral in some ways, as several employees left to start new companies.

Tommy Green, who worked at Knowsley, started Trophy, Vanroyce and later Vanmaster (no connection with the glass-reinforced plastic models). John and Cyril Darwin and Brian Talbot and Ken Wilcox (Lunar) also did their

After its initial launch the Swift range acquired its sleeker shape.

apprenticeships at Knowsley, as did the founder of Viscount. The company was basically responsible for the whole north Lancashire caravan manufacturing complex.

At Freckleton, near Preston, George Foley (ex-Pemberton) and Leslie Holland (ex-CI, Wilk and Astral) set up Fleetwind Caravans in 1967. From experience they knew they would be in direct competition with Sprite for sales (a daunting prospect since their 12ft model cost £7 more than the Alpine in 1969). The teardrop front roof design meant you could spot a Fleetwind fairly easily and, although basic, they did sell, though not at Sprite levels (at that time, though, nothing could really compete with Sprite).

Pemberton, which had by then withdrawn from producing touring caravans, took the step of announcing a 14ft four berth in 1967, priced at £340. Aiming to undercut the Sprite Musketeer (which it managed by just £7), the Pemberton Diamond didn't make the grade, due to cheap looks and poor finish: production ended in 1968.

Ernley Caravans, near Blackburn, was another static holiday caravan manufacturer that briefly dabbled in tourer production. In 1962 Ernley brought out a 14ft four berth before going into liquidation.

Colonial, Skyline, Dovedale and Summerdale had all built tourers in Blackpool (though Skyline produced only a few and stopped in 1963 after just a few years). Colonial launched its Texan ranges and had considerable success in export markets, although static holiday homes would take precedence at the Colonial Lytham factory. The Texan tourers became Olympics, and later became known as Olympic Caravans (now Stuart Longton Caravans Blackpool dealership). Colonial stopped production in February 1969.

Another Blackpool manufacturer making a brief appearance was Tancrest Shult. These tourers had a rather plain exterior profile. Ex-Colonial boss Bob Greenhaigh was involved and tried to attack Sprite's market on price. Apart from the little rear-doored 8ft Shult Cub, the range was a flop. Quality wasn't good, even though bonded sides were innovative at that time, and twelve months later, in February 1970, the four tourer range ceased production.

Summerdale, over at Thornton, near Cleveleys, dropped the Summerdale name with tourers and produced the Sport range in 1964. These were relatively normal run of the mill tourers, except for the long distinctive side windows which allowed plenty of natural light and ventilation. Summerdale fell by the wayside after a takeover deal fell through in 1967.

Dovedale's tourer range finished by 1967, after Pemberton took over the company in that year. Dovedale's tourers did look a little dated by this time, although a notable model in the range was the Minum, a tiny 8ft 6in tourer, which had proved an export market success. Manchester-based Lynton's tourers were sent to Germany as well as being relatively successful on the home market. Names such as Valetta, Venita, Vega and Venus did okay in export markets, but didn't have the same impact at home. They were well-equipped tourers for the time, with the washroom in the17ft Venus having a tiled-effect wall and a washbasin.

Lynton wanted a larger slice of the UK tourer market. Reg Dean, who had established himself as a designer at Caravans International, was approached by Lynton management in 1966 and he took up their job offer. He was responsible for the classic Lynton swept-back roof profile. Developing ideas he had at CI, Dean completely revamped the Lynton tourers and the holiday homes, too.

For 1967, the swept-back roof profile was an instantly recognisable feature, and a peacock blue central band gave the range a sporty look. Interiors were bright and modern, with white pvc-faced walls and ceilings, blue upholstery and orange curtains. The carpet was red with black fleck, the furniture was teak veneer and the worktops were melamine-faced. Dean's attention to detail was obvious in the neat stainless steel cupboard door handles. It all sounds rather garish by today's standards, but was a hit with caravanners then.

Scamp, Arrow and Javelin were new model names which sounded as exciting as the Lynton tourers looked. The layout of the Arrow included a central washroom in its four berth 12ft 10in body shell. Dean had started a design trend with this layout, which every caravan manufacturer would eventually use, and the layout is as popular today as it ever was. The 14ft 10in Javelin featured four berths with an L-shaped front kitchen and corner washroom. A central settee offside bunk, with a dresser opposite and a rear dinette, were other Dean layout ideas. Dean turned Lynton's fortunes around in twelve months, increasing its market share considerably.

Inevitably, some manufacturers from the 50s wouldn't survive far into the next decade.

Marfleet's Sovereign tourers (incidentally, the author's parent's first caravan) became part of Mardon in 1971. Box-like in profile, the Sovereign's shape of 67/68 was given a makeover for 1969.

Even though there was a boom, it didn't stop some makers going out of business, such as Berkeley, early in 1961. Berkeley had set some pioneering trends over the twelve years it had been in business However, involvement with the car industry (car enthusiasts will remember the Berkeley sports car), the car's costly development and failed American export sales killed off both caravans and cars. With debts of £332,000, and a take over bid by another company falling through, Berkeley went into liquidation. Panter, the firm's founder, set about making a comeback in 1962 with the help of Geoff Holden of Paladin.

Called the Encore, the new Berkeley van was basically a reborn Delight, a glass fibre model from 1956 based on 14ft body lengths. Unfortunately, it didn't take off as Panter thought, and even with the help of a 22ft holiday home, sales were disappointing, and the Berkeley name went out of production again.

Paladin and Normandie, who had both seen tourers sell well in the late 50s, went over to holiday home production by the early 60s. The tourers that were built were primarily for export, though also available in the UK. Normandie built the A models until it went out of business while Paladin built the 425 until the end of 1963. For 1967 Paladin launched its only tourer, the Stowaway, a 10ft four berth with a difference. As its name suggested, it was easy to store, in fact, the whole van collapsed to a flat pack (very MFI). It didn't catch on, unsurprisingly, and was Paladin's final tourer (caravan production came to a full stop in 1967, thereafter only special units were built). Donnington, the Newbury manufacturer, also concentrated on holiday homes but built tourers, such as the Nine, until 1966. Its wrap-a-round rear windows gave an air of spaciousness to its 9ft length.

Harvington, at Eversham, stopped tourer production in 1962, going over to retail sales. Jubilee tourer production at Wednesbury, Staffs, was cut back to just two 14ft models (the Firefly, produced for just a short while didn't find favour and was omitted from the line-up). Jubilee now ceased building tourers and, instead, like Rollalong, changed to making specials.

Nene Valley, until then a staunch holiday home manufacturer, moved into tourer production for the new decade. Based in Rushden, Northants, the company produced its first true tourer in 1961, the 14ft Dart, which is worth mentioning because of its rear bedroom, unusual in a tourer of this size. It also featured a triple-panelled roof for added insulation. To complement the Dart, Nene Valley launched the Dove, a 10ft tourer. By 1965, however, Nene Valley Coachworks became part of the American corporation Divico Wayne, which also owned the Dutch touring caravan manufacturer, Kip. 1966 was the last year that Nene Valley produced tourers; thereafter the company concentrated on static holiday home manufacture.

Roy Cattell (Nene Valley founder) formed the breakaway Estuary Caravans at Felixstowe in Suffolk and launched his Cavalier tourers for 1967. These were box-shaped, with radiused corners and doors on the continental side (Sweden and Scandinavia being the biggest export markets for the Cavalier). The tourers

A-Line's super tourers had metallic blue lower body panels, making them distinguished-looking. This is the Super 12 from 1969.

gained a strong following among caravanners, with their continentally influenced layouts and interiors ("teak" veneer finish on interior wall panels, along with Afromosia furniture, proved popular). Another neat feature was the extractor hood over the kitchen.

Over in Harwich (having moved from Romford) the Hammerton brothers, Derek, Michael and Terry, were building tourers for the export market. Fairview mainly concentrated on holiday homes but went into building and adding more tourers by the early 1960s. Although tourer production was shortlived, the company did produce some smart-looking units: Amstel, Laagan and Indus made up the range. The 10ft Amstel boasted that its 9cwt unladen weight was within the towing capabilities of the Mini, and it was priced at £299 in 1964. The Laagan was a 12ft tourer with a drop-down double bed (still a common feature in tourers), and a small corner washroom which Fairview equipped with a "washbasin and toothbrush holder!" Quality control at Fairview had previously been a major problem, and several fires at the factory didn't help matters. Despite quality checks and regular meetings with staff and foremen, Fairview quality improved only slightly. At the Earls Court Show for the 1967 model year, only one tourer, the Riva, was displayed, and this was for export only. Fairview had, like many caravan manufacturers of the time, produced tourers for special requirements, even building mobile offices for the RAF. Later, Fairview concentrated on holiday and mobile homes until the company's demise in July 1971.

As the end of the decade approached, competition and demand meant that those manufacturers established pre-1960 had to make a more concerted effort to take advantage of this buoyant climate. L and A Fisher of Walton on Thames and Bagshot, Surrey, had been making holiday and mobile homes since the 50s. The Overlander had given the company good steady sales, with tourers being more of a sideline. Tourer production started in earnest in the 60s with the Holivan range. Fisher concentrated on small tourers (8ft in length) and their distinctive exterior had a quilted side panelling. Although glass-reinforced plastic was being used by some caravan manufacturers in the late 50s, this new decade witnessed it being used more extensively. For 1961, Fisher introduced an odd looking glass-reinforced plastic model called the Siesta. Its 8ft length provided two berths and it cost £195. This, though, was the only year it was produced, and it was Fisher's last attempt with glass-reinforced plastic. Fisher

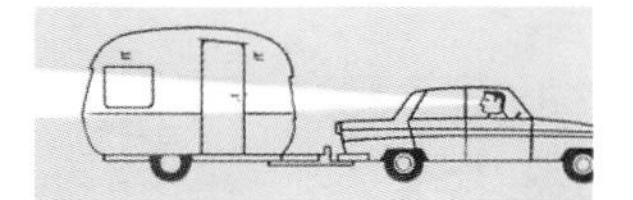

ROBIN FOURTEEN

14 ft. x 6 ft. 3 in. SLEEPING FIVE

BE RIGHT — BUY ROBIN

(ANOTHER OF OUR LUXURY RANGE)

Robin became one of the well known East Yorkshire makers, the fourteen model from 1969 could sleep five.

The interior of the Robin fourteen, like most mid–priced tourers of that time, had real wood veneer. Gas lamps, a two-burner grill and a foot pump were all basic essentials - then classed as standard.

From 1967, the Silverline Pullman Royal Lowline, available in five 12ft layouts, became known as the Pegasus. Silverline built up large export markets.

also made 12ft and 14ft tourers, but the tiny tourers proved more popular with small car owners, due to their light weight and narrow width.

Stephens and West, who produced the super-luxury Stirlings, introduced the Waterbird Cygnet in 1963, a new, small, lightweight range. Designed especially for the small car owner, its 9ft length was ideally suited to the Mini, and came fitted with Mini-sized wheels. Cygnets became Stephens and West's specialist compact tourers, produced in two and four berth layouts, and later in 12ft layouts. Profiles were simple (they came with flat roofs) and specification was basic, this being reflected in the price.

Down in Bristol, Bailey had grown at a steady pace, with exports playing an important role. In 1964 the home market had tourers such as the distinctive-looking Maru, a 10ft compact four berth. Bailey also manufactured the Mikado, a 12ft tourer, in various layouts. The long-running Maestro also received several alternative layouts. For 1968, the Bailey exterior design was given a much needed facelift, after the company was criticised by the caravan press for its dated-looking tourers.

Bailey's new look was a winner and the old image of sound but unexciting tourers was replaced by one of quality stylish tourers. Interiors were freshened up, with a new African veneer called Zebrano for the furniture. Bailey's popularity was just beginning to take off, slowly but surely.

Bailey's caravan manufacturing neighbour, Gold Star, founded in 1968, fell by the wayside in 1971. Rodway, another Bristol manufacturer, that also built boats, was a low profile manufacturer who made a small range of

Stephens & West was better known for Stirlings. The Cygnets, though, proved just as popular as lightweight tourers. The 12ft & 9ft from 1968 were best sellers for the company.

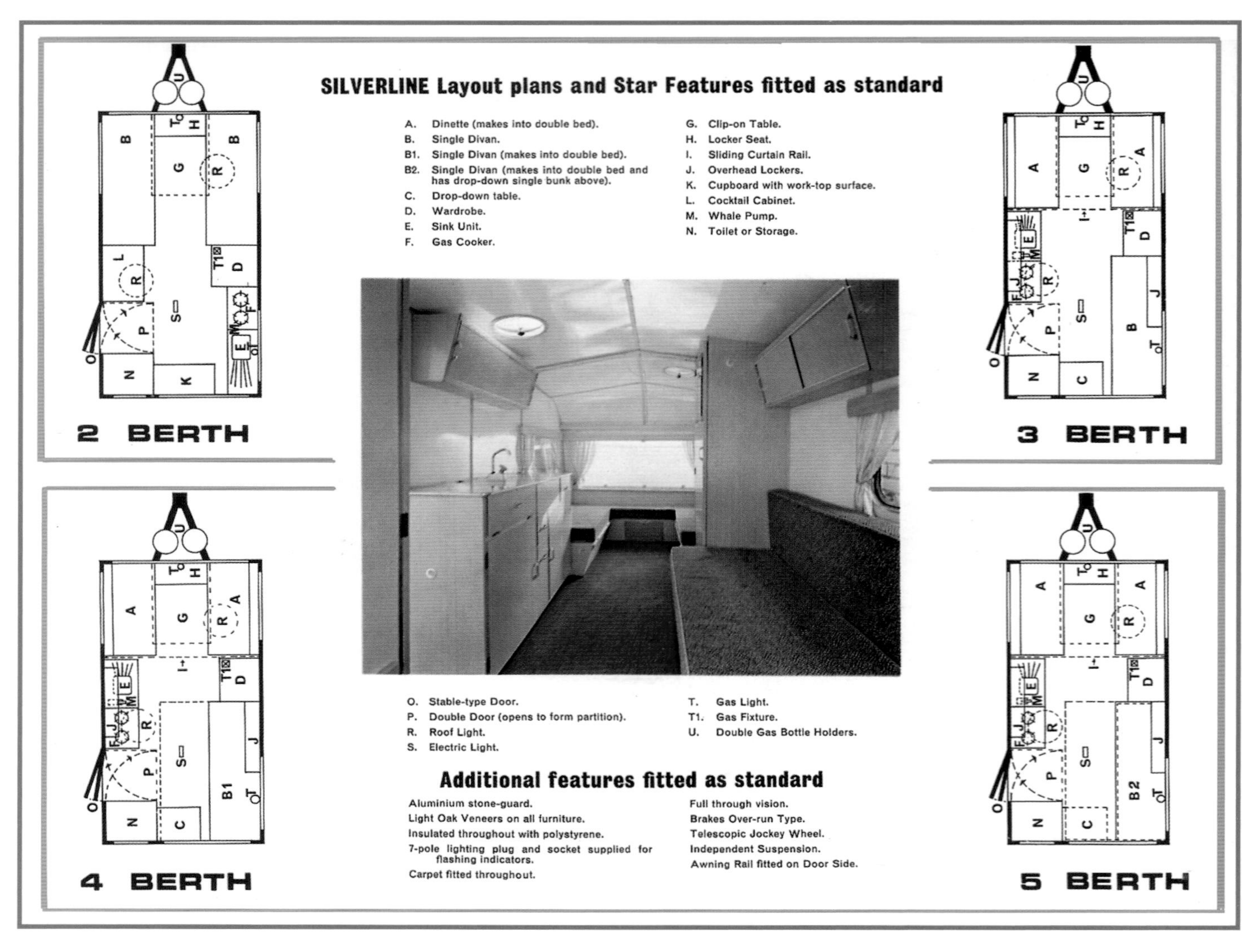

Pullman interiors had light oak veneer, hand-operated water pump and a gas point for an optional fire.

tourers. After 1970 the company built to order only, although they did reappear for a short time in the early 80s.

In Wiltshire another virtual unknown, Avon caravans, built a few old-fashioned looking tourers from 1968. These were mainly 10ft tourers, with a basic interior but a quality feel (the distinctive Avon could be easily recognised by its chunky profile).

Sutherland, who made the popular Shannon range, also made a 12ft tourer in 1963 called the Minor. The following year a 13ft 6in tourer called the Fleetwood was launched. It wouldn't be until 1967, though, that Fleetwood tourers saw the light of day again (with three models). That was the last year of production under the old Sutherland company name.

Harry Brown broke away from Sutherland and launched the new Fleetwood tourers from Colchester for 1968 (the Fleetwood Colchester models came onto the market place in 1969). Great success in export markets, mainly Sweden, saw the special continental range built for that market and for home consumption as well. The plain but practical UK range acquired a small but loyal customer base.

Amphibious tourers never made large sales impacts (the idea was a good one, in theory, but sales were far too few). The Amphibian Otter tourer, made in very small numbers from 1957 to 1966, was considered a novelty, but the Amphibian company was joined in this market by Glider, the old established Northants company. Apart from a Slipstream tourer range, Glider launched a 16ft amphibious model called the Carafloat. Glider finally stopped production

Reg Dean designed the stylistic swept-back roof of the Lynton tourers. Dean was also keen to promote his caravans; the above background was chosen to show off the 69 range.

in 1967 after 30 years.

Creighton, the Nelson-based company which made clubman tourers, also added an amphibious tourer to its range, called the Gull (probably the best-looking of all amphibious vans, though the purchaser had to pay an extra £100 for the outboard motor). The amphibious tourer sank without trace (pun intended) with only a brief attempt by a small maker, Caraboat in Mansfield, Notts, to refloat the idea.

Although this book is about UK makers, it's worth mentioning the attempts in this decade to launch imported brands. The German make, Tabberts, came close in 1967 with its well-built, high-spec tourers. Constructram, the Belgian manufacturer, also made a failed attempt at the UK market in 1964, as did Caravelair of France in 1966. Sterckeman came to our shores with an eight van range at the same time but didn't get the response it needed. All were interesting tourers but they weren't to the UK buyer's taste. Irish tourers, such as Pull-Van, Rolon, Sylva and Shannon, did enjoy some sales success here. Freedom (no connection with the Polish imported tourer) was distributed by Callender Caravans at Carnforth, Lancs, in 1967 and, out of all the aforementioned Irish manufacturers, was the most popular, setting up a network of several dealers before going into liquidation in 1970.

As we come to the close of this long, but important, chapter, we have seen how the British touring caravan industry became renowned for its choice and innovative skill at producing touring caravans that sold worldwide.

The last ten years, to 1970, saw the 50mm towball become standard. Towbar manufacturers, such as Witter and Dixon Bate, made towbars for all car designs. Most caravan manufacturers used an off-the-peg chassis such as that made by Peak or B&B Trailers, and specialist suppliers such as Carver, Whale, Hunter Aluminium, Lab-Craft, to name a few, helped give the industry a better image. Independent suspension, full road lights, awning rails, gas points, full insulation, better upholstery, fluorescent lights - all had come along to make the 60s tourer a better product. Sites, too, had become more civilised by way of wash and toilet facilities. Proper pitches were planned and levelled, making caravanning that much more enjoyable.

In the next chapter we will see how the clubman and super clubman manufacturers grew, including those in the vibrant Hull area. If you were caravanning in this period you may have heard of some of the manufacturers, or you may even have owned one of the tourers.

At the end of the 60s, the mainstream tourer manufacturers were on a high after the record-breaking Earls Court Show. Blackpool-based manufacturer Olympic took export orders for £53,750, equivalent to 125 tourers. Silverline, which had now moved to Full Sutton, near York, sold 350 units valued at £250,000. Willerby and Robin took £650,000, with the only cause for concern being whether the shipping lines were able to meet Willerby's

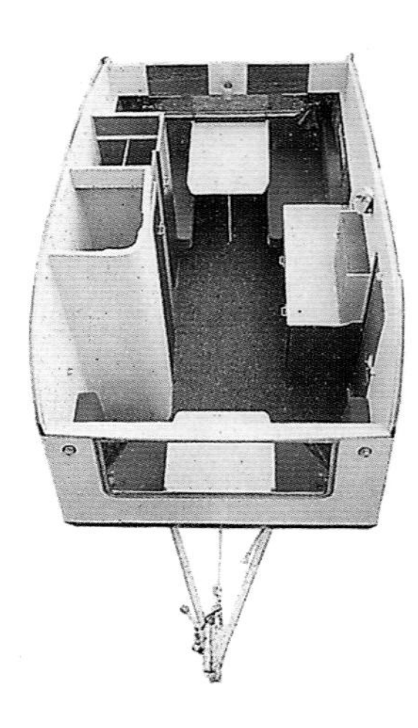

commitments. A-Line took orders for a total of 900 caravans, from tourers to holiday homes, worth £400,000, and established ten export markets. Industry giant, Caravans International, took an order from Scandinavia for just short of 6000 units for urgent delivery. The order was worth over £2,000,000. Sprite, Bluebird and Eccles were the main caravans wanted. CI now had 55% of the Danish market and 30% of both Norway and Scandinavia. The Earls Court Show in November 1968 pulled in 118,000 visitors, 1100 buyers (home and abroad) with sales totalling over £5 million.

The British touring caravan industry had finally come of age.

Reg Dean's interiors were another first. With his flair for producing off-the-wall designs, blending white vinyl with teak veneers, he created a spacious feel. They also came with specially-designed Pyrex Lynton crockery.

The rise of the Clubman luxury touring van 1960 - 1970

Sprite was dominating the ever-popular budget end of the touring caravan market, and Thomson was still a leading choice in the middle price range. The other two sectors, the clubman and the super luxury clubman tourers, were led by Fairholme/Eccles and Carlight respectively.

We've seen how the caravan industry reached new heights, was being recognised as an export earner, and created new supplier sources. The mass market had grown, and, in the same period of time (the 60s), sales of both the clubman and the super luxury clubman increased. As a result, the number of manufacturers eager to cater for the caravanning enthusiast who wanted the best rapidly grew. Cost didn't matter, although quality definitely did. Sometimes it was simply a case of oneupmanship. At the end of the decade, 30 dedicated manufacturers catered to this more discerning customer. Some of the old favourites, like Siddall and Car Cruiser, came to the end of production, but were later relaunched by ex-employees or other caravan-related companies.

The clubman tourer was a quality caravan which came with very few extras, whereas the super luxury tourer had a fridge, a heater, possibly a shower, and hot and cold water. Hull, surprise surprise, produced some of these tourers, so it is here that we will start.

Clubman manufacturer, Welton, situated in the village of the same name, initially produced only 12ft and 15ft models. By 1963, the well-known Welton model names Galliard, Rosaleda and Miravista were launched.

The Welton shape slowly evolved from what was a bland, outdated design to a neater profile, and in 1962 Welton introduced its special boat roof design (it was flattened instead of peaked). This gave the company a distinctive feature for its models which was soon copied by other manufacturers. Equipped with a front integral gas locker, as well as a rear enclosed water carrier compartment, it even had a lockable access door.

It was Welton's fine craftsmanship, however, which earnt the company such a

Buccaneer's first attempts at the clubman market was this 12ft tourer produced for the 1969 season. For 1970, the company completely redesigned exteriors and interiors.

Carlight was well-known in super clubman circles for unashamedly luxurious tourers. This is the Casetta from 1966; every caravanner aspired to owning a Carlight one day.

reputation for well-built touring caravans. The Welton shape had changed again by 1965, with this new shape continuing for a further nine years before being replaced. Although basic in equipment, the Welton was a sound tourer and gathered a strong following in clubman circles. Production was kept to around seven tourers a week, although the company also built special mobile units for industry and received orders from all over the world.

Hardly a drawbar's length from Silverline's Full Sutton factory (part of an old World War Two aerodrome) a new manufacturer, Buccaneer Caravans, was founded in late 1968 by DW Shipley (ex-Abbey and Lissett) joining up with NC Harrision, an enthusiastic caravanner. By 1969 the pair launched the first Buccaneer 12ft tourer. At a cost of £425, it came with a gas point, water pump and full polystyrene insulation. It looked like the Lissett Cuban of that model year; hardly surprising, given Shipley's background.

A complete rethink for 1970 saw the new and long-lasting Buccaneers come into being with Swift-like profiles and a "Welton-style roof." The Elan, the Caribbean and the Clipper were quick to follow.

Starting in late 1967, the year before Buccaneer, Embassy tourers set new standards by building a basic clubman van and then adding the "extras" as standard. Flued heater, hot water, fridge, water pump and stainless steel sink, all came as part of the Embassy package.

Harry Emms and John Tate, who had worked for Ace, knew that customers wanted value in this sector. Following their success they moved to larger premises, not far from Sovereign at Marfleet Lane, Hedon, and produced five layouts and sizes, from 11ft to 17ft. The Embassy in 1970 had a price tag of £888, making it excellent value.

Another Hull clubman manufacturer which began in January of 1969 was Royden. At Endyke Lane, Hull, Royden produced very nicely profiled tourers which should have sold on looks alone. Triple front and rear windows, radiused corners, as well as that Welton roof-line and low-key coachwork, made the 13ft De Luxe (priced at £575) quite a stunner. Within twelve months the company had added four more tourers to the range and moved to Walton Street, Hull. The tourers were packed with goodies, such as a full oven, on-board water tank, fridge, alarm clock, hot water, flued heater and 12 volt lighting. These were tourers which should have done well, but, unfortunately, production ceased in early 1970.

Still in Yorkshire, although not in the east of the county, Bessacarr, from nearby Rotherham, mainly produced mobile homes. By 1961, however, the company introduced its first real tourer range with the Royalette. With its easy to spot front profile (described as resembling a man's pushed-out chest), the Bessacarr drew much attention at the 1961 Earls Court Show.

Bessacarr's sapele finished woodwork could not be faulted, and this new clubman tourer was seen on many a rally site. At a later point, models were introduced which split the Bessacarrs into two ranges: the cheaper Continentals, with lower specification, and the Rally range, which for 1969, received quality glass-reinforced plastic moulded panels front and rear. Specification included a heater, a fridge and an oven.

Over in Ossett, West Yorkshire, a new clubman tourer, the Dalesman, was about to be launched by the co-proprietor of a construction company. Ken Shaw, an avid caravanner of 25 year's standing, built his own tourer and then tested it for three years before the first Dalesman tourers went on sale. Shaw used ideas taken from his aircraft construction days in designing and constructing the Dalesman. The van's framing was dowelled and glued at the joints instead of the usual half lapped and screwed method. For interior panelling the usual hardboard was rejected in favour of plywood, which was then glued to the framework, making a strong, lightweight construction. Another idea Shaw came up with was using just two sheets of horizontal aluminium panelling for the side walls. Not only was this pleasing to the eye, it also cut down the risk of leaks. Shaw kept the interiors uncluttered, almost along the same lines as Reg Dean's Lyntons. Eight tourers were in the Dalesman line-up for 1965, but by early 1966 Spen Coachbuilding (motorhome producers) had taken over the company.

Spen-Dalesman, as they were then known, offered double glazing as standard in 1967, something virtually unheard of at the time. Spen also added a glass-reinforced plastic roof. By May 1968, however, the company had stopped tourer production altogether.

Short-lived manufacturers were nothing new; the Compass Caravan Company in 1969 (no connection to the Compass of Explorer Group fame) was such an example. Fred Vowles, previously with Pearman Briggs (manufacturer of Safari caravans), and a Mr Alan Blick, entered the market with the 11ft 6in two berth Cub. Its lines resembled the Safari, complete with lantern roof. Built on a B&B chassis, the Cub weighed in at a hefty 13cwt unladen. Specification was basic, with a foot pump and gas fire point, although the quality was high. Very shortly afterwards, however, the company went out of business, not having lasted the year.

Within two years Fred Vowles had teamed-up with David Owen of Fairford caravans

The Dalesman was little known in clubman circles. Its design wasn't dissimilar to that of the Car Cruiser, and body construction was both strong and unique. This is the 10 from 1964.

Bessacarr produced one of the best-known clubman ranges of tourers, with a distinctive front end. The 12ft Rex, from 1964, was a four berth tourer.

Car Cruiser, while still respected in the luxury market, was slowly losing ground as new makers, such as Castleton, came on the scene. This Carousel, from 1962, was the largest Car Cruiser at 15ft 10in.

dealership in Gloucester to manufacture the 13ft Kimberley two berth. It was an odd-looking tourer which had a high specification, including a fridge, oven and heater, and an awning light. Interestingly, the Kimberley was equipped with separately operated hazard warning lights, using the van's own power source (a very innovative feature for its time). However, the Kimberley didn't last a season.

What we must remember at this point is that, although touring caravan sales were going well, with so many new makers, something had to give! Some (such as Siddall, the old established Cheltenham-based luxury clubman tourer manufacturer) got a new lease of life. Many were saddened when, in 1959, the company ceased caravan manufacture. However, in 1960 the make was revived when the then ever-expanding caravan distributors Gailey bought the names and moulds. Siddall was back. Gailey gave the job of building the new Siddalls to the Kelston Caravan Company, which was building mobile homes at the time. The 1961 Siddall Torbay, a full 17ft end-kitchen four berth, cost £1150 and weighed in at a whacking 27cwt unladen (nearly 32cwt laden), and all on a single axle! The glass-reinforced plastic roof and front and rear panels oozed charisma. Lime oak-finished furniture and soft luxury furnishings completed the luxury, up-market feel. Neat, innovative ideas included a side enclosed gas locker, an awning locker, tip-up washbasin in the washroom and a fitted chemical loo.

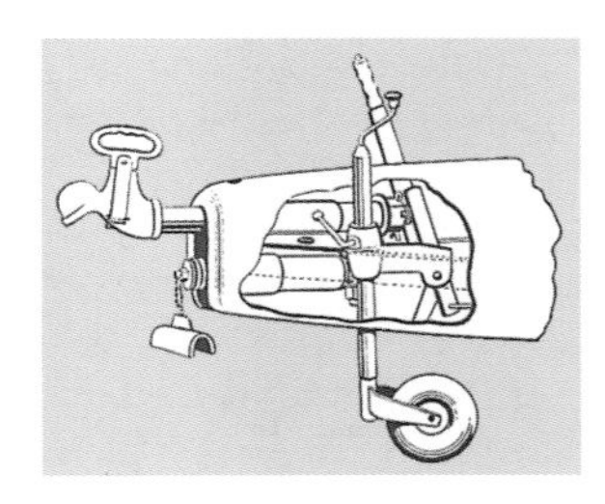

Gailey also proved that the Siddall could be a trendsetting tourer when the 15ft Delta came about in 1962, designed by industrial designers, the Conran Design Group. Tom Karren (later designer of the 1970 Eccles Amethyst) gave the Delta a pleasing rectangular shape. But, although it looked the part, it simply wasn't very practical. Its layout took in an open-plan design and included two washbasins, a washroom, gas heating and hot water, all for £545. Gailey stopped Siddall production for good in 1965, and out with Siddall went the Delta.

Although they manufactured more specialist caravans, *i.e.* larger showman models, Lonsdale at Culgaith, Penrith and Vickers also built heavy-duty clubman tourers. The Bantam Cock and Game Cock 12ft and 14ft models were produced by Lonsdale, whilst the Morecambe-situated Vickers made the Featherlite up until 1962 (a four berth 14ft costing £495). Vickers carried on making large showman vans after this.

Balmfourth Sanderson continued to make special clubman vans, built to order only, but this came to an end in 1967 (a great pity because the tourers were well made with years of craftsmanship behind them).

Vanmaster (no connection to the 1999 Vanmaster) built shapely, clubman type, glass-reinforced plastic shells. In 1968 Vanmaster teamed up with Chas Mitchell, the Avon caravan manufacturer, that, for a couple of years, fitted out Vanmaster as a clubman tourer, making them available as 12ft two or four berth layouts. Furniture was finished in a choice of light or limed oak. The newly fitted out Vanmasters were known as the Vanmaster Avons.

Just as Siddall had fallen and then restarted, so too did Winchester, maker of the "Rolls Royce of Caravans." The last Winchester, the 16ft Voyageur, was produced in 1959 and cost £1250. In 1962, however, the Winchester was given a new lease of life with the help of Stephens and West, maker of Stirling super luxury tourers, and the smaller and cheaper Waterbird Cygnets, just outside Cirencester. Stephens and West purchased the Winchester name and made the Winchester **Pipit (Pippit?)**, joined by the 15ft two berth Widgeon in 1963. The Winchester name was put into mothballs again after this, but was relaunched 10 years later for 1974.

At this point we move on to Stirling Caravans. At the turn of the 1960s, the graceful-looking Stirling was a heavy, coachbuilt, super luxury tourer. Specials included fittings such as a heater, an oven, venetian blinds and a fridge. Stirling built its own heavy-duty beam-axled chassis, although changed to a B&B unit later on.

Superb woodwork and quality fittings, and the large lantern roof design, made Stirlings popular well into the 70s. One dealer organised a Stirling test speed tow in Italy in 1965, and managed a speed of 100mph. Stirling, like so many of the clubman manufacturers, built special mobile units, with Stephens and West building mobile units for the Pathfinder caravan park. Stirling's profile changed very little in the years to come. Stirling also exported some models, a few of which went to America.

The Cheltenham, one of the most famous clubman tourers in UK caravan history, continued to sell well into the 60s The glass-reinforced plastic shell, with curved aluminium side panels was, and still is, classic. Gardener, the family that owned Cheltenham, had slowly induced their son, Cecil, to take on the business. For the 1961 season, the Puku 14ft two/four berth was introduced, making six models for that year. Its features included electrically pumped water, a shower, a fridge and an extractor fan.

The quality of woodwork and design saw the Cheltenhams at their most popular through the 60s. The most sought-after of the Cheltenhams were the Puku, the Sable and the little 11ft Fawn, designed for, and sold in, export markets. Cheltenhams enjoyed a loyal owners' club, with which the Gardener family was very involved, organising many weekend meets at the family residence in the Cotswolds.

Like Cheltenham, Holgate Caravans could, in time, have become another living legend. Holgate had always considered quality to be a key factor in the clubman market. The factory premises, at Great Harwood in Lancashire, had at one time been a cinema and it was from here that the clubman range of tourers was produced.

Attention to detail was one of the company's strengths and, although not widely known, it sold well. The Silver series of tourers were named Snipe, Wren, Wings, Cloud, Arrow and Silverlite, with the same exterior profile being used for many years. The range was modernised for the 63/64 model years but still managed to retain its clubman identity. Holgate stopped production in March 1964 and, although the Holgate family had planned to launch a new, cheaper range of tourers, they invested money and energy into their now very successful holiday home and touring park near Silverdale instead.

Creighton, over at Nelson in Lancashire, was building clubman tourers with odd model names such as the Poodle, the Deerhound, the Cairn and the Corgi. In 1967 it had an amphibian tourer (mentioned in the previous chapter) and the Lowline range, which were aimed at the mid-market. The Capri was the clubman model, a 16ft four berth with end kitchen costing £825. Its pleasant-looking lines and good specification were selling points, but Creighton designers had fitted an unusual sliding sunroof for summer touring. It must have been a real leak risk! Creighton stopped production in late 1967.

Located in an old disused woollen mill in

Holgate was still producing tourers up until March 1964. This is the Silver Arrow from 1960, a 16ft four berth model which weighed in at 18cwt. It featured a full L-shaped rear kitchen. Holgate tourers were well made but did not receive the respect they deserved.

The famous Cheltenham. This is a Sable, from 1962, being towed by a Triumph Herald. This picture was taken at the Cheltenham owners' club rally venue, the Gardeners' Southfield Farm in Gloucestershire .

the village of Heads Nook, near Carlisle, one of the best known clubman tourers was built. In 1965, Viking Fibreline Caravans emerged. The Viking was based on the Africaravan design, originating from South Africa, called the Gypsy. Glass-reinforced plastic moulds were used extensively in the Viking's construction, making it aerodynamic and very distinctive, as well as having incredible strength. All glass-reinforced plastic work was designed and built in-house.

The 1966 model year produced a very stark Viking (no painted panels and a lack of veneers and roof locker storage) but these were added in 1967 by Tony Holland, the company's owner. The Viking became an almost instant success, with good press reviews on its towing characteristics and build quality. Specification-wise it didn't boast much until the 1970s.

At one time Vikings were possibly the most exported of all the clubmans, and production was increased by the introduction of a night shift in 1967. A notable feature of the Viking was the glass-reinforced plastic entrance door which was made by the company. Leaks in Vikings were rare, due to the one-piece roof, front and rear.

The Sable's interior of the same year. Quality was what the Cheltenham - one of the most famous clubman makes (and the oldest) - was all about.

At about the same time as Viking began, a Birkenhead shopfitting company saw a market for a full glass-reinforced plastic clubman super luxury tourer, and in late 1965, at the Earls Court Show, launched the 1966 Regency range. Trade reaction dubbed the Regency a bit of an ugly duckling, what with its bulbous looks, especially on the shorter, 12ft models. With the glass-reinforced plastic body, Regency used a polyurethane foam filling sandwiched between two layers of plywood, an idea ahead of its time in terms of insulation and strength. No-one could argue that the GT models were not well equipped, since they featured a full shower, a Carver 1800 heater, a fridge, oven, hot and cold water, on-board water tank, fluorescent lighting and, on later model years, an electric brake system.

The 16ft 6in Monarch, with a price tag of £1085, came equipped with disc brakes and, on site, its easily distinguishable profile caught the eye. Teak veneer was used for all the furniture, with roof lockers actually being made from glass-reinforced plastic. Regency also built export specials, making them available for UK buyers as well. Regency was ahead of its time in many ways, even offering removable carpets and auto-reverse.

Ensor Caravans, the West Bromwich company, had carried on from the 50s into the 60s with little changes. Joe Bissell sold his Ensor tourers through, and on to, original Ensor owners. Ensor clubman tourers were noted for

The first Vikings, for 1966, were basically the same as the South African design. Within two years they became clubman orientated.

high quality among more serious enthusiasts. Fine cabinet work and Ensor's own patented draughtproof windows, glass-reinforced plastic body components and extra-long drawbar for more stable towing, were all Ensor features. Ensor models consisted of the Kiwi, the Eden, the Elf, the Merit and the Compact. Only one van a week was built and Ensors were rarely seen: they were probably one of the UK's lesser-known clubman caravans.

Car Cruiser had been one of the real founders of the caravan industry and had produced some classic tourers over the years, but had seen stiff competition from the "new kids on the block" such as Dalesman and Castleton. Blending modern design with traditional craftsmanship, its tourers were still held in high regard in clubman circles. Low in specification but high in quality and practicality, Car Cruiser had six models, ranging in price from the £425 lightweight Carissima to the £835 15ft Cambridge. A feature of the 1963 range was the B&B chassis which had rubber independent suspension, along with built-in jacking points.

In late 1962 the company moved from its old home at Hayes to a new factory in Cambridge, a move which caused panic in the market as old Car Cruiser buyers thought that quality would slip (as when Sprite acquired Eccles). Things didn't go smoothly but the problems that did arise were soon sorted out by the new owners, the Primrose Group. A major change for 1966 was that all new models were known by numbers. However, shortly after the new models were announced, the company was wound up.

The unthinkable had happened, Car Cruiser was no more. Ex-managing director, Bob Cooper, bought the name intent on bringing the company back; (Cooper, who had initially served his time with Berkeley, had moved to Car Cruiser to take on the sales operation). Alongside Cooper's specially-built units, the latest Car Cruisers were introduced. Cooper took some of the old ideas and rehashed them for his new 1969 four van layout. Car Cruiser vans were back for the time being, but, although quality was excellent, the vans were dropped, and then made to special order only.

Clubman tourer manufacturers were all fighting for a slice of the market and competition was now rather fierce. Even Irish maker Pull-Van, which produced a very solid and well equipped tourer along the lines of Safari, lasted only a couple of seasons. Shannon, based near Dublin, produced a nice-

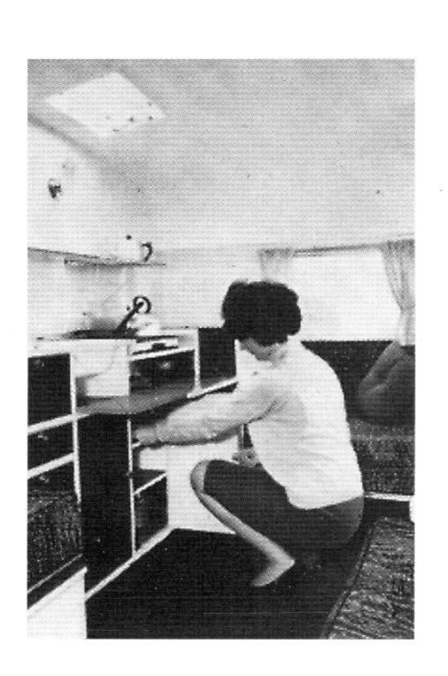

Welton was a small Yorkshire clubman maker, whose reputation for super quality woodwork went before it. This model, the "Twelve," would have set you back £435 in 1962.

A Welton Express pictured in the pretty village of Welton in 1970. Other makers, such as Swift, Bessacarr and Andromeda, also used this setting as a brochure photo backdrop.

looking clubman tourer but failed to make an impact. Although Irish manufacturers did export caravans to Holland, Germany and Belgium, most did not survive after the end of the 60s.

Established manufacturers such as Safari had a loyal following which, in common with many manufacturers, had established owners' clubs. Safari vans had a distinctive profile, used since the company began. In 1962, however, the shape was changed, although it could still be recognised as Safari and still sported a lantern roof and curved side windows (a feature unique to them). The new Safari was modern-looking and fairly well equipped (they came fitted with fridges) and, just like Cheltenham, Safari became one of the best-known clubman manufacturers, producing caravans that many aspired to own.

In 1968, Cosalt (the maker of Abbey) bought Safari to market as its luxury range. Safari was basically left to run as before, although up and coming Tony Hailey from Abbey helped with the transaction. Although Cosalt mainly produced tourers designed for couples, it also produced family layouts, with models designated by length, such as the 17/2 and the 14/4.

Down at Tinneys Lane in Sherborne, Dorset, Les Bennett was producing his Castleton clubman models, since starting the company almost by accident in 1958. His Castletons were good practical clubman tourers which, within several years, were giving Car Cruiser a bad time. The workmanship was always of the highest order, although specification-wise they did lag slightly behind their direct rivals. The Castleton profile was well founded by the mid-1960s and a reputation for durability, as well as an active owners' club, saw the make become increasingly popular.

Towards the end of the decade, a new super clubman tourer was launched, very much in the same format as the Regency. In fact, Karefree Leisure, at Bentley, Doncaster was a shopfitting company which, I think, must have been heavily influenced by Regency. Trading as Carapace, it introduced the Quartet 14ft tourer, which relied on glass-reinforced plastic moulding for both the front and rear panels. The Carapace had more style than the Regency and its exterior was more pleasing to the eye. Interiors had teak veneers in full use, changing to light oak the following year. Hot water, quality soft furnishings, a heater and a fridge were all standard on the Carapace. The company also exported its tourers, mainly to Holland. In a very short time Carapace had built a good strong dealer network and looked set to stay – for the time being at least.

Better known for mobile homes, Travelmaster, based in Caernarvonshire, also had a five year run at building clubman tourers, and offered the buyer good basic value. Named the Tasman and the Texan (not to be confused with Colonial), these tourers proved their worth with caravanners who wanted quality and a clubman image at a reasonable cost. Travelmaster listed model sizes from 11 to 17ft in 1968. They were easy to identify with their ever-so-slightly dated exterior. Travelmaster ended tourer production rather abruptly in late 1968, and from then on made only large mobile homes.

As we near the end of this magical era of the clubman and super clubman tourers, we come to the king of kings of super luxury touring caravans. Carlight Trailers had built up the Carlight name by improving quality, whilst at the same time making them the most expensive, and sought-after tourers in this market sector. Carlight owners, Bob Earl and

Interior of the Welton Rosaleda 13ft four berth model from 1970. Welton's fine craftsmanship can be seen here.

Castleton was winning many accolades for its clubman tourers. This Castleton Rovana two berth and Singer Vogue from 1966 won the Brighton Trophy.

Safari, another of the great club-class tourers. This is the 1966 15/4 complete with full lantern roof. Like the Cheltenham, a real design classic.

Cyril Gregory, had cleverly designed the Carlight's exterior, making it unique and very much a tourer for the connoisseur. In fact, the company built up a very good customer base in the Middle–East, where customers were very demanding, needing air-conditioning and insect screens specially fitted to the tourers. Meticulous attention to detail and customising were to be Carlight's key selling points.

Not ones to let new construction ideas pass by, Gregory, whose experience in glues and plastics was invaluable to the company, used his knowledge of one method which involved integrating the wooden floor and steel chassis, and gluing the inner panels to the framework (just as Shaw did with his Dalesman range). The result was a lightweight and strong design - a touring caravan manufacturer's dream.

In 1964 Carlight also used a sandwich construction for the side walls for its 14ft Cassetta. The outer skin was glass-reinforced plastic, melamine was used for the interior panelling and the core was expanded PVC. There were no seams, and it was double overlapped where it joined the front and end glass-reinforced plastic panels. What Carlight did 35 years ago, Abbey Caravans now use with its 1999 Evolution range.

Carlight had other interests, such as making and selling an auto-reverse kit and the Carlight patented towing stabiliser. One of Carlight's optional extras was the vacuum-operated braking system, available in 1969 at a cost of £14.

So how much did a Carlight cost? Well, the big 17ft Continental would have set you back a cool £1680 in 1967. That would have bought you four 16ft Sprite Majors in that same year!

In the next chapter we will see how clubman manufacturers hit trouble, and the caravan industry change as a whole. New names arrived, and some old ones went for good. With takeovers, and medium-priced manufacturers chasing new clubman markets, the 70s were certainly a time of change.

Travelmaster's Texan from 1965 was the biggest in a range of four clubman tourers. Skilled workmanship and good build quality brought only limited success.

The Creighton Deerhound from 1960 has clubman appeal but didn't make it in the popularity stakes.

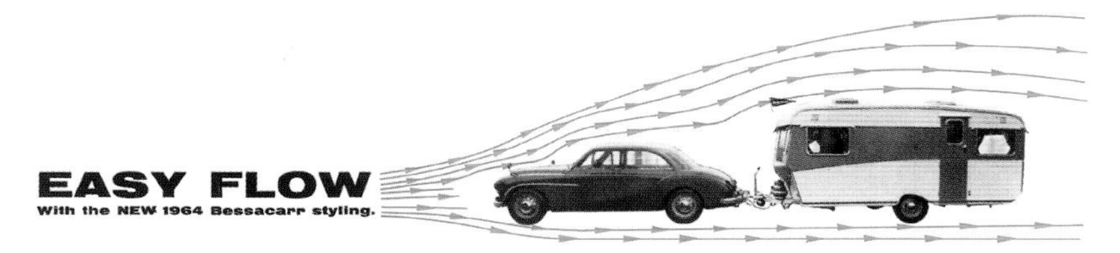

Regency interiors were very upmarket, especially by 1968. Craftsmanship and standard equipment supplied made them very special.

The Regency was one of the best-equipped super luxury tourers around. Glass-reinforced plastic was used extensively for the outer shell.

The pace-setting 70s

Growth in the 1960s - which gave the caravan industry new status - continued into the early part of the 70s (the caravan press stated in late 1969 that the "touring caravan explosion was all set to carry on").

Hull was still considered as one of the UK's main areas of caravan manufacture, second only to Newmarket. The touring caravan manufacturing industry was still very active and, at the start of the decade, several new names in tourers, some from established manufacturers, others from manufacturers new to touring caravans, surfaced.

The first headline news in the new decade was that motorhome manufacturer Dormobile was branching out into building touring caravans. The company had built its first trailer back in 1922 (a 17ft unit which, apparently, went to Argentina), but this latest venture would involve a new factory in Ashford, Kent. Dormobile was targetting the Ace/Swift/Astral market sector, and the new marketing manager (an ex-Maidstone caravan centre manager), had big plans for the three tourer line–up.

Plans were to produce 5000 units a year by the end of 1971, which would make Dormobile a major player along the lines of Ace and Astral, whose production and model ranges had

The Mustang Pinto from 1974 cost £378; its 14ft length and 4 berths proved popular. Bill Boasman, the company's founder, based the Mustang profile very much on the Abbey. Typical garish 70s colours were used for the Mustang interiors.

increased at rapid rates. Dormobile's plans ran into slight problems, quality-wise, even though the factory had modern equipment, including thermobonding presses and air compressed tools. Dormobile's expansion plans were impressive and ambitious, but proved that tourer production had to start off at a steady rate. Perhaps Dormobile thought tourer manufacture success would have been assured given the company's famous motorhomes, but this obviously wasn't the case. Quantity production came to a halt pretty much before it started. The following year the company started low production runs with a super-smart and distinctive-looking clubman tourer named the Kent.

The tourer factory was moved from Ashford to the company's Folkestone works. Dormobile stopped making tourers in 1973, although by 1977 they were back with bonded constructed and double glazed designs influenced by their continental buyers. These proved far more popular than the previous designs and, for a few years, Dormobile tourers were seen in reasonable numbers on roads and sites. Dormobile boasted in advertising literature that its caravans were suitable for all climates.

Tranby, part of the Hull-based Rootes car dealer Triangle Motors, distributor for Bluebird motorhomes in North and East Yorkshire, was, in 1970, the most recent clubman maker. The caravan manufacturing side of Triangle Motors was instigated by Ken Hennebry who had just moved back to the UK from New Zealand and had, for a short time, been involved with the production of Shamrock caravans. The four berth 13ft clubman tourer didn't have the much-loved triple front windows, or a lantern roof, and it was rather expensive (£640). Spec-

Silverline clubman interiors were described as richly appointed; this is the unusual side kitchen two berth clubman from 1972.

Astral launched the budget Scouts for 1974, dropping Rangers, which were brought back in 1975 by dealer demand. The Scouts remained in production until 1977.

SILVERLINE-SHOW STOPPER of the 70's

. . . illustrated here, at the 1970 Earls Court Exhibition, is possibly the first caravan in the world to be finished outside in nylon fibres. This process is not new, as it has been in use for about 25 years on hundreds of different applications throughout the world. You may like it, or hate it, but in your interest, we're busy planning ahead to the future with new ideas . . . new finishes . . . new designs . . . new concepts of caravan comfort for you and your family. Take a look through this leaflet — pay a visit to our distributor, and you'll discover just why Silverline is truly the gilt-edged caravan investment of the Seventies!

For recommended retail prices see separate price list. While the information contained in this brochure is correct at the time of going to press it may be that on account of the use of alternative materials or the shortage of supplies alterations may have to be made to the caravans described in this brochure. Particulars of any changes may be obtained from the manufacturer.

The Silverline Clubman GT, with its optional nylon fibre coated exterior, was a real showstopper - it never caught on, though.

wise it came with a fridge and a gas heater but the interior was a bit of a nonstarter, compared to the Buccaneer, anyway. Tranby wanted limited production - and got it - the company folded within a few months!

Eton, another dismal failure in the Hull area, announced a 12ft medium-priced tourer which bore a resemblance to the Abbey; Eton disappeared at around the same time as did Tranby.

A new manufacturer, based at Beverley, not far from Riviera at Grove Hill, was Beverley Coachcraft. Ex-A-Line chief Tony Warburton had a new £40,000 manufacturing plant built for the three van line-up. Looking very much like Estuary's Cavalier tourers, with an almost box-shape profile, the Beverley had more pronounced radiused curves, and also looked better.

Sandwich construction, based very much on the way tourers are constructed now, was used on the Beverley's side walls. Interiors, verysimilar to those of Elddis, used Japanese oak veneer furniture with orange upholstery. In 1971, Beverley Coachcraft was building "just-towable" 22ft tourers, and it was three of these models which were bought by the *Blue Peter* programme with an "old knives and forks" appeal campaign to raise money for three caravans for under-privileged children. Beverley had a couple of minutes' exposure on a prime time TV slot.

Minster was another maker that set its sights on becoming one of East Yorkshire's top tourer manufacturers. Formed as the 1970 model year came in, the company was founded by ex-Astral sales manager Craven Giplin. He put his tourer factory into production at Victoria

Silverline entered the new decade with this Clubman GT from 1972.

Rivieras were medium-priced tourers with an easily recognisable shape. This is a1972 14ft Rimini five berth model.

Road, Beverley. Minster's Huntsman four van range was later augmented by the addition of a holiday home. Minster was geared-up and ready for exports, and all models had doors on the offside, even for the UK market.

Some tourer manufacturers started by building touring caravans but eventually concentrated on holiday homes. Castle was one of these, producing the 13ft Windsor, a triple front windowed tourer, with distinctive grey top and lower panels. The Castle company was founded by ex-Royden team Mr C Davis and Stan Berriman, at Leonard Street, Beverley. Within four years the company had moved on to holiday homes only. Aaro also went down this path after it stopped production of the 13ft Primella tourer in 1972.

Elddis, for a long time the North East's only touring caravan manufacturer, was joined in the field for a short time by three new manufacturers. Sturdiluxe, whose factory was in Chester-Le-Street, was an offshoot of H Young (Motors) coachbuilders and motor distributors. Company executive, John Hedges launched the 13ft Weardale in 1970. Its exterior profile was plain and relatively uninspiring, although a new two-colour nylon finish was tried on the inside: "sunshine orange" for the wall panelling and olive green for the furniture. It all sounded dreadful and *Caravan Magazine* testers also noted that the nylon fibre dropped off easily! This type of finish was also used by Silverline on the exterior where it was offered as an optional extra at the 1971 Earls Court Show.

During the following twelve months, export-orientated models joined the Weardale, although the velvet nylon finish wasn't adopted for them. By May 1971, Sturdiluxe was no longer producing tourers.

Carville Caravans was launched in May 1970 by Darlington coachbuilder MT Service Co at its Victoria Street premises. Commercial body-builders by trade, as well as trailer and cattle truck manufacturers, the company also built the occasional motorhome to special order. The founder of the caravan division, Mr R Tefler, a long-standing caravanner, based his 12 and 14ft models very much on the Ace shape of 1968. Carville vans were sold only through the company's retail premises and were shortlived (by 1972 production had stopped).

In Sunderland, Sinclair-Gordon produced two super luxury clubman tourers, with a shape reminiscent of the Safari, and intended for the clubman market. Alan Brewer, a bespoke cabinet maker, joined forces with David Grabham, who owned several garages, and began tourer manufacture. Grabham had actually built and raced a car constructed mainly out of bonded wood, proving its strength.

The pair came up with two touring caravan models. As well as the design, the construction was also unique. Two skins of 4mm weatherproofed ply were bonded to the hardwood framing, with the aluminium skin over this. The construction was so strong that seven men could stand on the roof without it showing any sign of weakness. This method of construction was used for the early Vanroyce tourers sixteen years later. The Sinclair tourer should have been a winner, even at £850, but by 1972 production suddenly stopped.

The medium-priced tourer market experienced stiff competition as new makers tried to muscle in. Casualties would be inevitable, and this made for exciting and anxious times, as tourer manufacturers jostled to become established and quickly make a

Foot Pump

Push Button Handles

Royales were super luxury tourers. The heavy use of glass-reinforced plastic is evident and looks its best on this 1977 18ft Tourcruiser, with a hefty 26cwt unladen weight. Inset: The interior of the Royale Tourcruiser is opulent, with a high standard of equipment.

name for themselves.

Another new maker in 1970 was SBQ (Smith Bros. Quinton) a small woodwork and prefabricated building company based at Oldbury, Worcestershire. The director, Mr Pugh, was a keen clubman caravanner who attended many rally weekends in a 14ft tourer he had produced himself. Feedback regarding

Avondale tourers were a clubman's delight, offering good specification at a moderate price. A Rover and 1974 Mayfly two berth are perfectly matched in this Avondale sales shot.

Interior of the Globetrotter was practical but well made.

The popularity of Ace caravans grew: the Globetrotter pictured here is a 1971 model, a favourite with families.

this prototype resulted in launch of the SBQ 14ft Debonair and the 12ft Nymph. Their looks were heavily influenced by Ace, and, to the untrained eye, could easily have been mistaken as such. The much-favoured oak veneer was chosen for the furniture and full insulation, a water pump and a gas fire point were included as standard. Distributed through Gailey, Dunsley and Fairford caravan dealers, the SBQ met with limited success and folded in September 1971.

It wasn't all doom and gloom, however, since, not far from Knowsley caravans, would soon begin one of the most successful names in the industry. The name chosen by Brian Talbot and Ken Wilcox, ex-Knowsley employees, for their new tourer manufacturing venture was Lunar, after the Lunar space programme in 1969. Under pressure, after being sacked by Knowsley boss Edmund Taylor (who found out they were planning to go into tourer manufacture), the pair had no option but to continue plans to go it alone, before completely ready to do so. They began to manufacture tourers in a somewhat unlikely place, an old barn near Ince, Wigan. Amazingly, with no heating or lighting, the first 10ft Lunar two berth, named the Saturn and priced at £398, was finally finished (after what seemed like weeks!).

The first dealer for Lunar was Barney Campbell, who had himself only just established a dealership at an old railway yard at Lockstock Hall in Preston. Within a few months Lunar was officially launched and the range was increased to include the Jupiter and Venus models. Lunars had more than a hint of Knowsley about them, including the drop–down sink unit for end kitchen models, but within a few years they had established their own identity.

Ski caravans was a small tourer manufacturer set up near Leigh in Wigan - also set up by ex-Knowsley staff. Very like Lynton in exterior design, the Ski tourer enjoyed little in the way of success and stopped in 1975, after only a four year run.

In 1972 two more Knowsley employees, Tommy Green and John Darwin, left to go to

Fleetwood produced this neat-looking profile in 1971, changing it very little for the coming years. The 1300 four/five berth was the best-selling Colchester.

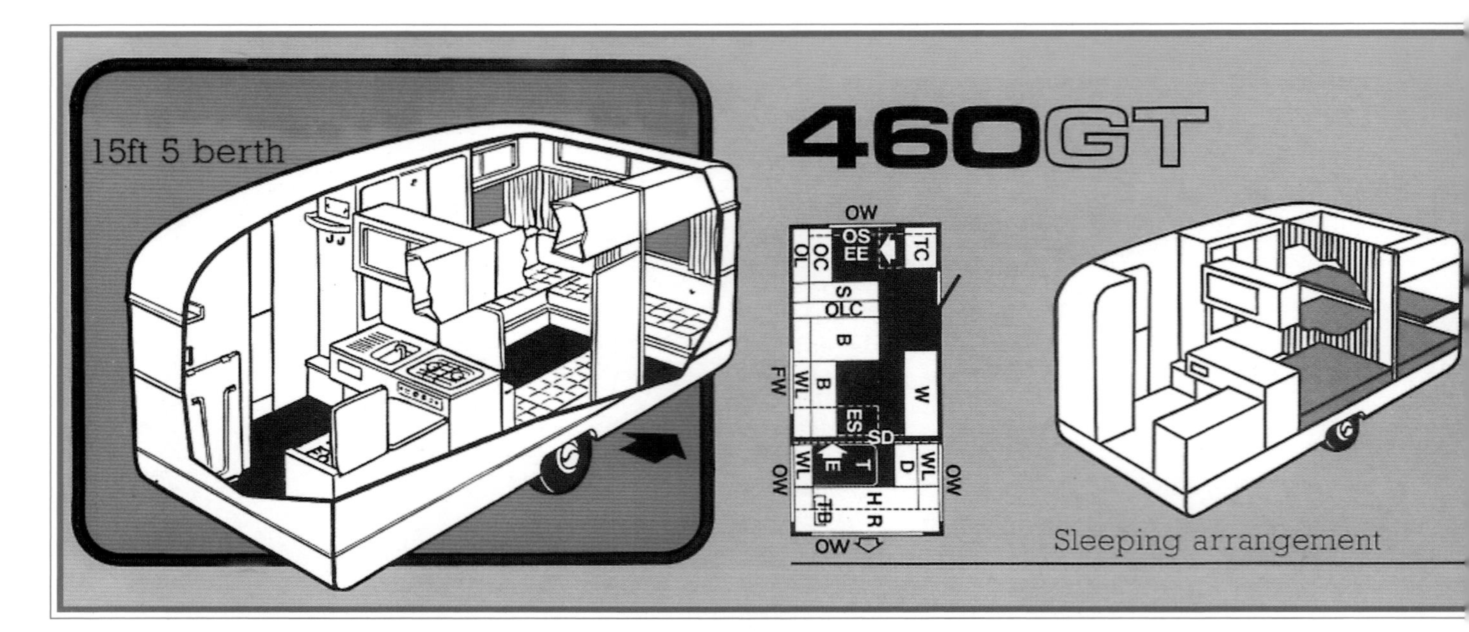

This cut-away shows how the 460GT looked, with its interesting rear L shaped layout, which was not dissimilar to Lunar's 1999 Lexon.

Liverpool to help start Trophy (which acquired a reputation for specialist lightweight tourer manufacture). Moving premises shortly after, to a converted mill at Croston, Trophy extended its range, and the plain-looking profile was altered. Basic interiors became more functional than luxurious, but the competitive price kept Trophy on the road of volume sales, especially with the flat-roofed Silver range, which looked very much like Pipers, plus the clubman Gold Trophy, which had more than a touch of Lunar.

Less than twenty miles away, in Freckelton, Fleetwind was making statics and tourers, mainly for export. Priced at just above other budget tourers, the Fleetwind vans found a niche. In 1972 the company launched a budget version; the Kestrel, a very basic van built in three sizes, was a little dubious with regard to build quality, with one magazine noticing that the side wall "flexed" when leant against. Despite this, Kestrels eventually outsold Fleetwinds.

Gerald Ball, an active caravanner himself, decided he could design and build a value-for-money clubman tourer. Ball had been Rootes Cars' chief industrial engineer at Coventry, and had also served five years at Jaguar Cars as a work study executive director. He launched his new clubman range for the 1971 model year. The Avondale tourer was unveiled to trade and press and received an excellent reception. The two models, the Kingfisher and the Mayfly, were super looking and well-equipped 13ft clubman tourers with two or four berths. Easily recognisable exteriors and luxury interiors of excellent craftsmanship, combined with a reasonable price tag (£765), made the models an instant sales success. After a very short time the Avondale owners' club was formed. Changes for the 1972 model year included a glass-reinforced plastic roof and a new moulded rear panel. Avondale had become a very serious threat to the other established clubman makers.

In September 1973 Ball saw a new opening for budget tourers and launched his basic Perle range. Costing £750, the 12ft van came with a drawbar cover, a gas bottle carrier and 12 volt lighting. Its plain but good looks won over many first time caravanners and it quickly became established, soon increasing the model line-up. A major breakthrough came in the industry when, in late 1973, B&B Trailers announced its Sigma auto reverse coupling, which meant not having to get out of the car to disengage the reverse catch, making reversing an easier task. 12 volt water pumps, along with 12 volt lighting, were other features that were standard on most tourers by the middle 70s.

Royale, the then new super luxury clubman maker down in Gloucester, which had been formed by ex-Safari employees John Fudge, Gwyn Jenkins and Harold Woodward, created a tourer in 1970 which lived up to its prestigious name in every way. Heavy use of glass-reinforced plastic for the roof and front and rear panels, gave the Royale instant appeal. These well built, but heavy, luxury tourers demanded a large towcar, even for the smallest model.

Innovative features included slatted bedding lockers, slatted bed base and integral gas locker with internal access. It's only in the last few years that caravan manufacturers have used these ideas on the latest models. Another innovative idea was an underfloor spare wheel compartment. Avondales have used this feature as a unique selling point for several years now, but Royale got there first.

North Star, producer of small luxury tourers in Preston, was an offshoot of dealer Dave Barrons, and was founded by Edmund Taylor (Knowsley's founder) Ken French, and Barrons' owner Gordon Hold. A glass-reinforced plastic roof, front and rear panels gave the North Star GT a distinctive look to complement a high specification. Four vans a week were planned but it soon became a case of special orders only, building to customers' special requirements.

After Fairview caravans went into liquidation in 1971, the company of Hammerton Bros. came back into business in 1973 with the FC Tourer. This oddball-looking clubman tourer was made out of glass-reinforced plastic and was rather unkindly described, by the caravan press, as resembling a loaf of bread! Its bowed sides and overhanging roof line made it too distinctive, reducing its appeal. Interiors featured mock rosewood veneer-finished furniture, and, in the GT425/two berth, a permanent L-shaped front dinette, an idea few manufacturers copied. Hammerton also supplied the FC in shell form for fitting-out by the owner.

The Victor Mini tourer, built by Bradford dealer Kenmore Caravans, didn't need such a hefty towcar, due to its ultra small dimensions. Chasing Fishers' market, the small Victor was only 9ft long and 5ft wide and was available in both flat and peaked roofed versions. The little tourers came with full insulation, gas point and a fire extinguisher. For 1975 two basic 11ft and 13ft tourers were introduced but manufacture stopped in the same year.

At Saltburn, Cleveland in 1974, two brothers made caravans and motorhomes. The Margrove touring caravans consisted of two models which were easily identifiable by the

Bailey's long-running Maestro from 1973; this is the "M" model four/five berth.

By 1972 Lynton interiors had become very "loud," with bright 70s colours like this garish purple.

split front window. The owners had a park and also did repairs, custom building and caravan transport. High Peak Caravan Centre, Derbyshire, built the Comanche range of mid-priced tourers from 1974. Their plain interiors, though, meant limited sales so numbers made and sold were small. Columbian luxury tourers started in 1973 in Chesterfield, were very unusual in shape and looked dated. They were well equipped but were built to order only.

Roy Cattell, the ex-Nene Valley/Estuary Cavalier boss, now owned a timber company in Suffolk called Greens of Brandon. With John Amis and Ray Cook, also from Cavalier, he started the mid-priced Panther tourer range. Looking a bit like a shapely Cavalier, doors were also automatically on the "wrong" side for easy export production. Panther shot to tourer fame and a strong dealer network was soon set up. Panther, along with Bailey and Cavalier, was one of the first companies to use twin axles on mass-produced tourers. Godfrey Pratt, who had been responsible for the Avondale Perle range, left to work for Panther, redesigning the profile with more aggressive styling and, at the same time, introducing the better-equipped Panther Rio range. Panther later became part of holiday home manufacturer Sunseeker.

In Nottingham, poultry company Jacksdale entered the caravan manufacturing industry with an unusual-looking tourer. Forest was the company's new name, and the 12ft Sherwood the first model. Designed by a consultant architect, the Sherwood's front and rear sloping ends had the window deeply inset. It gave the van character and a distinguishable styling feature. It was not one of the best-looking vans around, but it was certainly a new approach. Makers such as Elmwood, Dean and Brentwood soon followed with similar designs across the model range. Forest went into holiday home manufacture for a time, mainly for exports, when tourer sales dried up in 1976, and although the company did make a come-back, it was for the 1977 season only.

In late 1970, Bill Boasman left Cosalt's Abbey tourer division to set up his own caravan manufacturing business. Mustang was launched, and competed head-on with Abbey, in some cases undercutting it on price by £20. Made not far from Abbey, at Immingham, Grimsby, the 1971 three van range was aptly named after horses. Mustang timed it right, and these Abbey lookalikes tempted new customers. Mustang was set to take on the UK market, building up healthy export sales in the process.

Lynton, like everyone else, joined in the very "hairy" Caravan Road Rallies. A quick razz round the race track usually dealt the fatal "reduced to matchwood" blow.

ABI's Monza broke new ground in the budget market, taking much needed sales from Sprite. Pictured here is the 1977 1600.

Thomson had lost its market lead by 1976, and was no longer a major player. The 1976 range consisted of the Clan (budget) left, Glen, centre and River series, right.

Buyers were attracted to Mardon's neat-looking tourers. This exterior profile was established in 1967. The model here is the 11ft four berth with loo compartment.

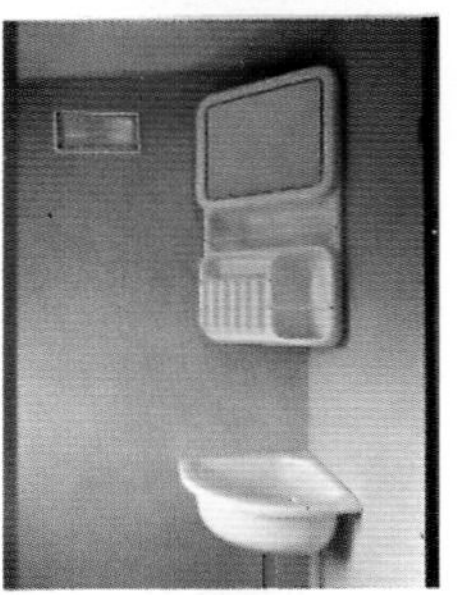

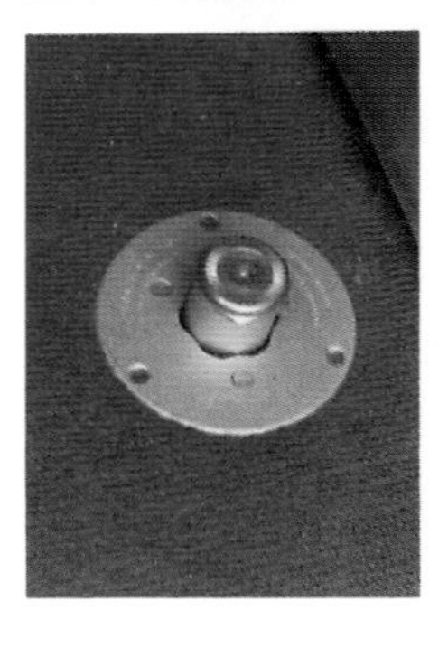

True to the caravan industry's tradition of employees leaving one company to start another, three of Boasman's employees left Mustang to start up their own concern - Boomerang - not far from the Mustang works at Grimsby. These wedged-shaped tourers boasted some interesting layouts; one even used the van's wardrobe base for the extension of a double bed. This involved sleeping with your feet in the wardrobe base!

The South Humberside caravan industry was completed when Jim Pearman, who had left Cosalt in 1972, started what became a famous clubman make - Cotswold Coachcraft. The specification in 1973 included a fridge, undersealed B&B chassis and full underfloor insulation. Typically clubman in looks and quality, like Avondale and Royale, Cotswold soon had its own thriving owners' club established.

Ace was doing as well as ever with exports and at home, although the previous 1969 model year was almost a disaster. That year was the first time Ace had used glass-reinforced plastic for the front and rear panels and quality became an issue as rumours circulated to the effect that the moulds, made by subsidiary company Ace Plastics at Ace Caravans' old premises in Clough Road, caused the panels to leak. These rumours were quickly denied by Ace chief, Terry Reed and, for 1970, the Ace range was completely redesigned using all aluminium panelling which distinguished the range and won over new admirers. It is still uncertain whether those glass-reinforced plastic panels caused problems.

Like Ace, Astral had become very popular, and had not only won a Queen's Award for exports, but also claimed to be the largest of the Hull caravan manufacturers. The Rangers, now easily distinguishable by the front bay triple window, had originally been built with the doors on the offside, but this changed in 1971 as home market needs were catered for. Astral launched the popular 10ft four berth Apollo for 1970, at around £20 less than the Sprite 400. Astral followed this initial Apollo with 12 and 14ft models for 1972 at a slightly cheaper price than the Rangers; however the vans didn't reappear for 1973.

Swift, in the meantime, had moved from the original factory at Heddon Road to Dunswell Road, Cottingham, in August 1970. This automatically increased production floor space to 15,000 sq.ft. Annual output of eight models was now around the 600 unit mark. Swift's body

In 1973, Knowsley introduced Bofors plastic Swedish double glazed windows, a first for mass-produced tourers. The Juno four berth had been stretched an extra foot to 13ft for that same year. Knowsley vans were at their most popular at this time.

profile had become established, and the low, triple, front and rear windows were a sales feature, along with the underfloor food storage compartment. Swift's original body styling lasted well into the next decade, such was its popularity. Swift, though, didn't follow other makers down the budget range road, happily selling plenty of the normal Swift range. Nobody could have guessed then that the company would become the UK's largest caravan manufacturer over the next 20 years.

Buccaneer, which had entered the clubman market the previous year with little success, deleted its two models. For 1970, a drastic styling revision was obvious with the all-new 13ft 6in Caribbean, which cost £595. Its design was very Swift, and the roof was the - soon to be well copied - Welton double peak design. A fridge, convector heater and fluorescent lights were all standard fittings. The Rubery Owen chassis with torsion bar suspension was dropped and replaced with the Peak unit, a favourite amongst manufacturers.

Dealers signed up for the smart-looking Buccaneer franchise, and two of the first were Barrons and Lowdham. Thirty years on, both

Interior of the 1977 ABI Target, followed Dean's way of thinking about touring caravan design.

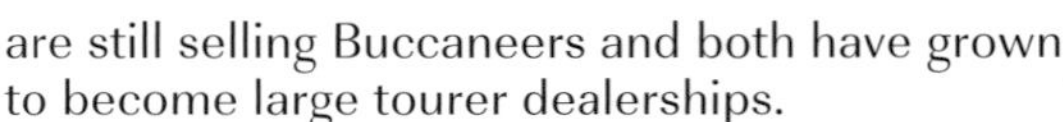

are still selling Buccaneers and both have grown to become large tourer dealerships.

In late 1972, Beverley was the home of Trinity tourers, produced for the super luxury clubman class. First impressions were that they looked like Astral's Ranger. The Trinity contained gas and electric lighting, mains electrics with charger unit, battery box and a complete list of standard items. Oak veneers ruled supreme. A larger model generated a small amount of interest, but, by 1975, had gone.

The old clubman make Vanmaster was now only available as a shell, as were Wake and Sapphire (also glass-reinforced plastic-based units). DIY tourers became unfashionable, though, and, by the end of the 70s, no more tourers of this ilk were made.

Fleetwood's Colchester range were smart-looking tourers, their neat, rounded corner body shells having instant appeal. Launched for 1971, they carried on in this format until the early 80s. Sweden and Denmark were countries which took on the specially-prepared Colchester range. The company left its Colchester-based works and moved to a new factory in the village of Long Melford, naming its export tourers Melford, after the village.

Production for touring caravans in 1971/2 was approaching 45,000 per year. Thomson, which was incredibly large now, was producing the Glen range, with a record output of over 6000 units a year adding to the figure above. Thomsons peaked in 1972, and, from that year

The ABI Target 3.90 was hailed as a modern-looking tourer. It was, of course, another of Reg Dean's creations.

on, began to lose ground to newer manufacturers. The company did try marketing a 12ft luxury tourer called the Clansman (on a limited two year production run). Based on the Glendale two berth, in 1973 it came with a fridge and a heater as standard.

In Hull a major surprise was about to be announced. Ace Caravans' founder, Terry Reed, held talks in February 1972 with holiday home manufacturer Belmont Caravans (established in 1966). The two companies were very close neighbours. To help expand both companies, just as Sprite and Bluebird had done over ten years earlier, they agreed on a merger and formed Ace Belmont International (ABI). This move resulted in a new force in the industry: the new company shot straight to number two in the UK tourer manufacturing charts, next to CI. The following year ABI took the opportunity to acquire the expanding Elddis Caravans concern, whose distinctive and smooth body lines, and generally well put together interiors, were in demand. ABI sensibly left the Delves Lane factory at Consett producing the Elddis tourers.

Inflation, fuel prices and the introduction of VAT saw the cost of touring caravans begin to spiral. The caravan industry began to suffer, with sales beginning to slow. ABI, Astral, Bailey, Forest and Abbey saw the Sprite budget market as a prime target for increased sales. ABI launched the Monza for 1973. Its sleek and sporty good looks, and build quality, offered exceptional value. Basically, it didn't feel like a budget tourer at all.

Monza sales figures proved that ABI was right to enter this market, and others quickly followed. Sprite, in its defence, added the £50 Club pack, offering a foot-operated water pump, carpet, stainless steel sink, gas locker and fluorescent light, plus Club graphics.

In 1973, Sprite celebrated 25 years in the industry. CI launched the limited edition Alpine model, with a few extras added, plus white exterior panels and silver graphics to celebrate the occasion. Just over 100 were built, one for each dealer.

Astral dropped all its Rangers and made the decision to launch the budget Scout, with

Russett Curtain

Russett Bedding

Lavender Curtain

Lavender Bedding

its very simple interior and exterior. Astral buyers didn't go for it, they wanted a Ranger, so Astral re-launched the Ranger for the 1975 model year, continuing Scout models alongside.

Bailey brought out the Prima midway through 1973. Its cheaper profile found favour and ran alongside the Bailey range. The Prima had a ready market in France, but cancelled exports meant that not many left these shores. Instead, Primas were found as bulk buys at some dealers. Abbey released the Pipers, which were based on the export orientated flat-roofed Abbeys. Although not as popular as the ABI Monzas, they still found a niche in the CI Sprite market. Just having the edge on its new competitors, it came equipped with a water pump, fluorescent light, stainless steel sink and fitted carpet, all of which was classed as good standard equipment. Pipers, although generally well built, did have some niggly quality problems, though, such as cheap window stays, and a few rough and ready edges.

Forest produced the Excel models which had a basic design along with a basic spec. The Excel didn't last for more than twelve months and Forest carried on with its normal, mid-priced tourers. Even the famous Cheltenham tourers acquired a cheaper range, called Explorer, a complete departure from the usual tourers, for 1972. The rectangular shape was softened by the glass-reinforced plastic moulded front and rear panels. The Explorer suffered from quality problems, a criticism which was levelled at Cheltenham tourers generally.

Knowsley, in a bid for sales, developed the tiny Athena in 1972 (later called the 290) which was joined by the four berth 390 (both were basic simple budget tourers). Knowsley clubman models were given a higher spec generally, including Bofors plastic double glazed windows as standard – a first for a UK medium-priced tourer. Knowsley UK tourers were built for Scandinavian weather conditions, with the Carver SB1800 flued heater as an option and all-round polystyrene insulation.

Lister, in Soham, had built caravans for years on a limited basis. It wasn't until 1972/3 that the company (renamed Mereside Coachworks) launched its super clubman tourers. These well equipped tourers provided the clubman with everything needed on tour. Enthusiasts looking for a distinctive clubman tourer put the Lister on their shortlist. Four years later, in 1976, the company had gone out of business.

VAT, and the general poor economic climate, resulted in large price increases and a slump in sales. The price of a Sprite Musketeer jumped by £144 in two years; Knowsley's Juno increased by £372; the Cavalier 1200CT by £218 and a Viking Fibreline 11B by as much as £321. This economic setback did hammer some of the smaller makers, with the larger ones feeling the pinch, too. In 1974 the power crisis hit and, with the threat of petrol rationing, touring caravan manufacture looked a bleak proposition. A poor year end meant that workers were, in some cases, being laid off.

Production was down by 40% across the board. CI had to lay off 180 workers at the Newmarket plant when dealers reported too much 1973 stock left over. Les and Alex Fisher of Fisher tourers had a three-day working week in force (they sold the company in May that year after briefly considering holiday home manufacture again). Cumbria-based Viking cut production, though thirty vans left the works in the first two months. On a footnote, Viking claimed its vans smooth, rounded corners helped save fuel, but during a fuel crisis it would say that wouldn't it.

Abbey, at Grimsby, had a generator in operation, which meant it had power for machinery and lighting during power cuts. This enabled Abbey to continue production an average of four days a week. ABI was struggling to produce anything, whilst Cygnet, Stirling and Winchester (reintroduced for 1974) were working flat-out on a three-day week.

Dealers, such as Goodalls (Countrywide Group) in Huddersfield, reported little in the way of falling sales, though, and touring caravan dealer Godfrey Davis saw sales slowly increase, due it said, to the package holiday offering less value for money. Years later, Godfrey Davis finished with tourer sales altogether.

In 1974, Cheltenham ceased production as did Explorer. Welton, which had re-vamped its range to include a heavy, old-fashioned lantern roof model for 1975, also went out of business. Stephens and West bought the Cheltenham moulds, and went into small batch production at the Cygnet/Stirling works. Production of the Cheltenham at the Stephens and West plant was slow and soon came to a halt again. Caravan dealer Ferndens, in Kent, took over production, with ex-Cheltenham kingpin Roger Launder, who produced a small number of caravans with the emphasis on quality and specification.

Even with this industry downturn, new

manufacturers still appeared and the industry as a whole seemed more positive. Humberside Caravans, founded by ex-Swift personnel, launched three stylish clubman-looking tourers, from its works at Grovehill, Beverley. Mallard, a small Northampton company, produced a 13ft van for 1974 only.

At this point we must not forget the "foreign invasion," as continental makers tried to infiltrate the UK market. Limited sales, though, kept imports at low levels, with perhaps the early 80s being the only period when there was a large influx of foreign makes. Adria made the biggest impact as a tourer import from 1971, equipping its tourers with standard items such as double glazing and mains electrics, long before UK manufacturers. Prices were also more competitive. Tabbert, the German manufacturer, had had three failed attempts on the UK market, even being imported by Abbey manufacturer Cosalt for a short time. MKP, the glass-reinforced plastic Danish tourer, was distinctive but too expensive, with prices double that of home producers; sales were limited (at £1970 for a 1973 MKP 14ft van, you could have bought a Carlight). Cabby, the Swedish maker, introduced well-insulated and well-built caravans, which had a spec to make even our severest winters tolerable, but the UK industry had this area all sewn up.

Back now to the home producers. CI, in a bid to gain dealer loyalty with the Sprite, Eccles, Fairholme and Europa brands, made a controversial move. All CI dealers were given

The Cavalier was one of the best-known tourers in the 70s. The GT versions were launched in 1971. The 460GT pictured here behind the "Wedge Princess" is a 1977 model.

the choice of specialising in just CI's vans, though they could include a luxury brand such as Safari or Carlight as well. The intention was to provide dealer support, i.e. advertising, refurbishment, promotion and a dedicated spares back-up, along with a CI finance plan. Most took up the idea, such was CI's influence within the industry, although CI's market slowly began to shrink. To counteract this, CI produced a brand new budget-beating tourer. Called the Sprint, this £700 12 footer was basic, very basic in fact. Its main selling points were price, weight (a shade over 8cwt unladen) and the dealer back-up. It was a hit, out-selling even the Sprite Alpine at one stage. At last, a Mini owner could tow a family caravan.

Sweeping changes at CI meant the Eccles range got a new shape and sandwich construction. The Eccles of 1976 went on for a further ten years with only minor changes, sales of it kept CI firmly on the touring caravan map. Even Sprite got a new shape, based loosely on the Sprints, and the market witnessed the return of the Aerial, which replaced the 400. Europas received more equipment, including a fridge for 1979, along with the Eccles' injected foam sidewall construction.

Fairholme quality had begun to slip, so, in an attempt to give the brand a new clubman following, a new factory was built in 1977 over at Bury St. Edmunds. CI added more standard equipment in an effort to improve the brand. CI had the new plant operated by six-man teams which followed the van through all stages of production, ensuring a personal touch and full quality control throughout.

By 1976, a real revival was going on in the industry. ABI had Reg Dean in its design and marketing team. After leaving Lynton, Dean turned his attention to the new ABI Target range, extending the ideas he had at Lynton. The Targets, with their wedge-shaped bodies, modern interiors and standard fitted fridge, were *the* talking point at Earls Court.

Earls Court also saw the new upmarket Elddis Crusader range launched, a really strong clubman market contender. A glass-reinforced plastic shell gave it the looks of a super modern tourer while providing clubman appeal and tradition.

ABI launched its first Award 12/2 for 1976, mainly as a limited edition Rallyman, which had won an award in a recent caravan road rally event. Public reaction meant the Ace Award became a separate clubman range for the 1977 model year. Abbey came up with the 13ft GT, which, at £1648 included double glazing, a heater, a fridge and special graphics. Swift, too, broke away from the medium clubman market by adding a luxury model, the high-spec 14ft Corniche. It proved a top seller so Swift increased the range. Mustang added a deluxe range, with the Arabian and the Palomino, which included a fridge, an oven and a heater to upgrade the specification. Lunar introduced the Clubman and Knowsley ran both standard and upgraded models, and adopted a super modern-looking shell for 1975, carrying on into 1976.

Instead of high specification tourers, Thomson introduced its budget Clan and River series, alongside its cutdown Glens. Royale launched a lower priced clubman van named the Windsor, making them at Beverley along with, in 1977, the Fiesta, an even cheaper range. At the Royale factory in Gloucester, a new super clubman range, called the Coronet, was added. New manufacturers, such as Lancashire-based Viscount, became known for quality and good practical clubman tourers. Cavalier launched its controversial Dart range as a down-market Cavalier. These were dropped, though, and replaced by the lower spec Cavalier Sceptres.

Lynton downgraded the Lynton shape in 1974, calling its new range the Lyncraft (after its boat-building involvement in 1962). For 1978 Lynton introduced the super luxury Trident, based on the, now defunct company which produced Listers. Astral was one of the first major manufacturers to use a new German chassis by Al-Ko for its export orientated Shadow range. A few years later most manufacturers followed Astral's lead.

The Welton name was revived for 1977, having been bought by Cosalt a few years after it purchased Riviera. Old Welton model names were brought back, along with some new ones.

Reg Dean left ABI and joined A-Line, where he carried on doing what he was good at, redesigning the range. A-Line and Dean then launched the Crown, followed by the Rambler and the Golden Crown.

More and more new names joined the industry at the end of the decade and, as this chapter comes to a close, we will take a brief look at some of them.

Cabrera, a Rotherham-based manufacturer, produced what were essentially Bessacarr clones. Elite Coachcraft in Bridlington was another company to make clubman tourers. Portman in Ipswich, ex-Cavalier people, set up a

Fairholme's luxury clubman image had become tarnished. This 1978 Fairholme Curlew two berth was produced at the new Bury St Edmunds factory in an effort to improve quality.

high-spec continental clubman tourer operation in 1977, while Chad, a small Bradford-based company, produced continentally-influenced tourers for a short while. Cougars, a dead ringer for Swift if ever there was one, were built not far from the old Colonial works near Lytham in Lancashire, for a short while, putting the resort back on the caravan manufacturing map. Reed Caravans took over from Victor Mini tourers by producing two small flat-roofed vans. Trio Trailers, in Hull, took the clubman market as its goal with the distinctive-looking Marlborough range.

Marathon, produced by ex-Avondale Perle man Godfrey Pratt, looked very much like a Lunar and made a brief appearance on the clubman market towards the end of the decade. EKP made Fisher-type tourers which were small in both length and width. Then we had the offshoot of Fisher, Cozycar in 1978, again chasing the Fisher market with an almost identically styled design. Monolite, launched for 1977, was formed by Derrick Rose in Gwent. Being an ex-Bailey man he had learned plenty from the Bristol-based company (which, incidentally, had been acquired by the Howard family). Monolites used a bonded construction technique, an aluminium chassis, double glazed polyplastics windows and 12 volt lighting. It was a very advanced tourer for its time.

Although Knowsley went out of business in 1976, the name reappeared, this time under the Pemberton banner. Knowsley tourers were in production again, built at the old Dovedale factory at Marton, Blackpool. However, this reincarnation was rather dull compared to the sporty 1976 version.

After sitting on the ABI board, Ray and Siddle Cook decided to go it alone again, setting up not far from Elddis, at Lanchester, Co Durham. The new Compass range was announced in late 1978. Citizen and

Commodore were listed for 1979 and both ranges caught caravanners' imaginations with their cosy interiors and plain good looks. Compass became a runaway success and the Cooks were now in direct competition with Elddis tourers, once their own company.

Just a few months earlier, Eddie Harrison and Ron Flowers left Swift to start their own tourer manufacturing company. Churchill clubman tourers soon gained a reputation for being solid, well-made tourers with a specification to match. Churchill's career had several stops and starts, with the company going into liquidation several times (this also happened with Monolite). Although Churchills didn't look dated, they didn't look as modern as the Alpha Autosport which came from Eric Prue's drawing board in 1978.

The Alpha broke all the rules, with its pronounced wedge-shaped frontal area and curved rear. Its specially designed Al-Ko chassis gave the Alpha excellent road manners. This was partly due to the fact that the axle was set further back than on conventional touring caravans. This design feature did have a drawback, in the way it would cut into corners on tow. Stability was its key factor, enabling the Alpha to break a caravan towing speed record in the early 80s. The Alpha certainly looked the part but quality, price, and a dodgy interior meant only around 130 units were built.

Inr a blaze of publicity at Earls Court, Avondale announced its export influenced tourer, the new Leda, for 1979. The caravan proved an excellent seller for the Warwickshire-based company, which received a further boost following the announcement that all Avondale clubman tourers were to be bonded constructed.

Although the latest batch of names had swelled the ranks of caravan manufacturers, it wasn't to last. With the recession of the 1980s on its way, the industry recorded its worst sales figures since inception back in 1919. In a last-ditch attempt to boost flagging sales, Astral launched new ranges which included the budget Flytes, the medium-priced Embassy and the luxury Cameos. It was hoped that these would fend off the fierce competition from ABI and Swift.

By the end of this decade, the touring caravan had developed into a leisure vehicle which was better equipped than ever (with heaters and hot water systems in vans such as Stirlings, Safaris, Carlights and Viking). Even medium-priced vans now had a fridge and plastic tinted windows. Caravans International introduced its Supercare scheme (later reincarnated by Swift as Supersure) as a unique aftersales service.

The British Caravan Road Rally was just about coming to the end of its life (writing-off brand new tourers wasn't as inexpensive as it used to be). Most were bashed into unrecognisable heaps of matchwood. In 1976 Silverline was presented with the Queen's Award for its outstanding export achievements.

Just as CI had done, ABI announced it wanted only ABI brands on dealer forecourts. The policy, however, was doomed to fail before it really got off the ground and was soon abandoned. The 70s - the golden decade of caravan manufacture - had seen the industry expand and decline. Although some tourers were a little dubious in quality, others were excellent examples of quality and design.

The 80s & 90s - recession and boom

The beginning of the 1980s heralded a new type of touring caravan, one which was light enough to be towed by the new generation of lighter cars, at the same time providing comfort for a new and more demanding touring caravan user. The end of the 70s was the last time there was a real "boom" in tourer sales. The 80s brought a slow-down in new sales and several manufacturers and dealers were bought out, or went out of business altogether (Caravans International collapsed, Al-Ko took over B&B Trailers, and Reg Dean rejoined ABI). Clubs had record memberships, though, and many sites were upgraded.

Cosalt bravely tried a "Tourer Design" competition in which it was hoped, the winning design would go into production. This came from a lady teacher in Manchester but when the van was built the poor reaction it received meant it became a one-off. In the mid–80s Cosalt employed Anne Parker, a local interior designer of soft furnishings, to do all tourer interiors. Cosalt also closed down the Safari factory at Stroud, in an effort to consolidate tourer manufacture, and moved it to Grimsby, launching a wedge-shaped Safari to boot, which was to mark the end of an era.

The start of dealer-built specials occurred around this time, when Harringtons and ABI got together to build a Monza with some special extras. White Arches did the same with Cosalt and the Abbey Acclaim, renamed the Archway. Gloucester dealer Ken Stephens had its own make, the Caraline. Swift expanded while other makers didn't, although the shrewd Smith family saw its empire expand even though sales were generally depressed.

Fleetwood Caravans became part of Swedish manufacturer Kabe, and Silverline made a comeback with its tiny luxury tourer, the Nova.

An important innovation at the time was the Tyron safety band, designed to keep the caravan under control if a blowout occurred. By 1989, the cassette loo was fitted as standard (or at the very least an optional extra) by most tourer manufacturers.

From the start of the 1980 season, as sales slowed right down, it was obvious that things were going to get tight. Too many tourers were chasing too few customers, and with over-production and the economy in the grip of recession, things were looking bleak. Even so, it didn't stop some imported makes trying for a foothold in the UK market. In the early 80s 3000-plus foreign tourers were imported, names like Digue, De Roeck (Jumatt), Laika and Polar, as well as Adria, Tabbert, Cabby, Hobby, TEC, Homecar, Fendit and Buerstner, were all trying to establish a market here.

UK exports slumped as continental manufacturers managed to hold their own on home territory. British caravan manufacturers found that drying-up exports could spell disaster, and even result in closure. In some cases this sudden collapse in exports had a very quick impact. In the first half of 1978, the market was worth £8.7 million (6294 tourers)

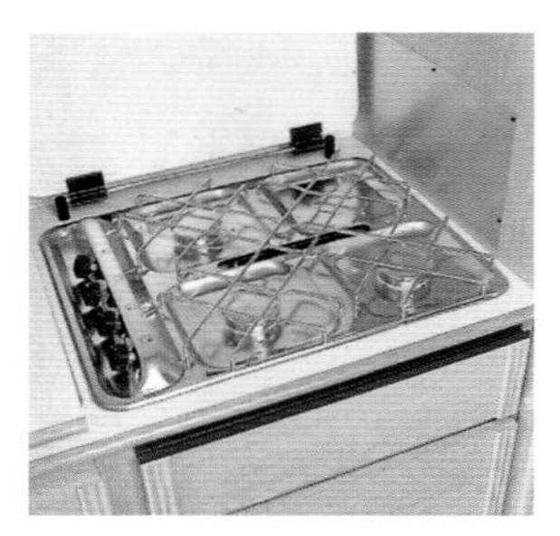

compared with £6.1 million (4255 tourers) for the same period a year later, a sure indication that things in the industry were tough.

Just as news of the 1980 Minster tourers was about to be released, director Craven Giplin announced that "difficult trading conditions" had forced Minster to cease trading. The giant Caravans International, whose market share had shrunk, especially in the Sprite budget area, launched the Cadet. Old name, old values; CI was trying to offer value, economy and light weight, whilst at the same time providing a family-sized tourer. The original 10ft 1979 model was joined by a 12ft version with centre washroom. An unusual feature was the Cadet's kitchen unit which was twin-bowled.

Reg Dean at A-Line had hotly contested the 10ft Cadet with the launch of his 10ft Imp in 1979. To compete with the Cadet 12, Dean launched an Imp 12. The cheeky Imp looked just like its competitor in profile, although inside was more homely and modern. Retail price was close, though the 10ft Cadet was a few pounds cheaper than the equivalent Imp. CI's factory had almost come to a standstill in September 1979 when a strike hit production for a month, postponing deliveries to dealers of the 1980 model.

Astral, which had improved its range no end with a total revamp for 1979, carried on for 1980 with detail changes only. The Cameo was a particularly successful van for the company, although the Flyte and the new Astral Embassy didn't do as well (not even with Embassy founder John Tate at the helm. He eventually went on to market Robin tourers at Willerby). It was soon evident, though, that Astral was facing hard times. The pound was strong and Astral depended on exports. In March 1980, the company laid-off its 90 or so workers and the factory compound was cleared (all 1980 tourers were out at dealers). The Spooner Group decided that caravans were no longer profitable and the factory was shut. Ironically enough, the old Stoneferry works were bought by Cosalt in later years, to make holiday homes.

Fisher, the small Surrey-based company, saw sales figures slide and it was eventually forced out of business. Although the Hull factory was sold off, Fisher started again, this time as GX Caravans. The new company produced some very unusual-looking tourers, although the Slimtreckker models looked much like the old Fishers. Some models featured double doors and even upstairs bedrooms.

Elite, the luxury Bridlington make, launched its 1980 models with some interesting features, one being a wind generator for charging the on-board battery. By July 1980, though, Elite too had faltered.

With weight and equipment being the main consideration, tourer manufacturers, such as Fleetwood, were changing over to the new German supplied Al-Ko chassis. B&B Trailers was making its taperline wishbone designed unit to compete on weight. Bailey was the only maker to use the new B&B HE aluminium chassis, on its M and Clifton range of tourers, though the standard steel unit was used for the base Baileys.

Peak Trailers had stopped chassis production, so many tourer manufacturers took on the new Ambergate galvanised chassis instead. Lunar went one better by designing and manufacturing its own lightweight aluminium chassis and glass-reinforced plastic panels, while at the same time setting up a new company called TW Chassis, who also supplied Buccaneer for a time. Cosalt's Abbeys chose the B&B taperline units for 1981 models.

The advent of the new bonded construction technique (aluminium exterior glued to the framework, with a polystyrene core and glued interior wall panels) allowed manufacturers to create lightweight yet strong tourers. The insulation properties were also superior to conventional glass fibre matting, which had a tendency to drop over the years. A lightweight chassis, combined with the new bonded walls, could make for a weight-saving of over 1.5cwt. This was further helped by the use of sandwich floors (except that some manufacturers' early models found the floor becoming spongy under foot as the material was breaking up).

Just about all manufacturers were by now including fridges and double glazing in medium-priced models. Elddis added a GT tag to create an up-rated Elddis caravan, which included features such as a fridge, a heater, spot lamps and hot water, supplied by the then new Carver Cascade (gas only) water heater. Showers were another feature only available in luxury clubman tourers a few years previous. The modern touring caravan was becoming better equipped as the decade progressed.

Cavalier Caravans, whose tourers had proved popular in the 70s, now began to lose popularity. The almost square profile with radiused corners was starting to look rather dated. In an effort to retain the familiar and recognisable shape, whilst at the same time

Cosalt's Abbey range of tourers for 1980 included the County, GT, Piper and Piper Clubman.

making it more modern, the front wall was made to slope back slightly. Cavalier's dominance had waned, and in 1982 it was taken over by Fleetwood. Willerby redesigned its Robin tourer range, giving them flat roof continental exteriors for 1980. After a poor season, however, Willerby launched the much shapelier Riband and a brand new twin axle model called the Argyll.

The Argyll was almost an exact copy of the ill-fated Astral Cameo, hardly surprising, really, given that Willerby purchased the Al-Ko chassis for the Cameo from Astral. Willerby ended its involvement with tourers when the three model Robin range came to an end in 1983.

At this point it is worth taking a look at the beginnings of the twin axle tourer as a viable unit. There have always been twin axle tourers, but these were basically built as specials. Cavalier built the 490GT on a twin axled B&B chassis and, a few years previously, in 1971, the 490-GT Bailey used a twin axle on the 19ft Mikouri. In 1980 ABI used the Target 5.70 as its first twin axle tourer. This new model came about due to the direct influence of the special Target built for motorcycle champion Barry Sheene in 1979. He had asked for twin axles to enable fast and stable continental towing.

Mardon followed suit, in 1981, with the Ultra 550s. At just short of £6000, this 18ft twin axle tourer boasted an impressive specification, including blown air heating and an extractor hood for the kitchen.

Competition was tough at this time and about to become tougher. Reg Dean's A-Line produced a super tourer, the Link, which became the company's flagship tourer range. By 1981 it was even equipped with electrically-operated hydraulic legs.

Luxury tourer manufacturer Safari stuck to its traditional shape, but added more in the way of luxury fittings. Stirling was making its heavy clubman tourers pretty much as it had always done. In 1981, Stirling reintroduced the Winchester; basically a more modern-looking Stirling, though of superb craftsmanship. Royale, part of the Aldington caravan retail group, was also still making super clubman luxury tourers. Although Royale changed the mouldings slightly, and generally spruced them up, they still bore strong Royale characteristics.

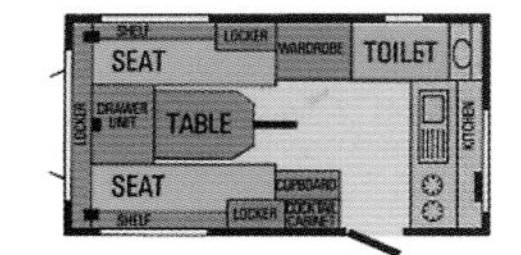

Lunar's Clubman 500/4 from 1983 was just one of a five model range, and one of the company's best sellers. The glass-reinforced plastic roof was made in-house, as was the chassis.

The company launched its most luxurious tourer to date for 1981, a £12,865 18ft model called the Tourcruiser which beat the famous Carlight on price. Carlight decided to go "modern" by revamping its shell (the first time it had been changed in 35 years). It retained its market position while at the same time giving off a superior air of quality. Cotswold introduced a new glass-reinforced plastic roof, which added an updated appearance, while Buccaneer and Castleton played safe with few changes.

Bessacarr bought the Astral Cameo moulds, which turned out to be a very wise move because, as every twin axle touring caravanner knows, the Cameo became a successful range for the company (eventually replacing all other Bessacarr ranges). In fact, it was the Cameo that really established twin axles as a firm favourite with the caravanner.

The clubman market saw a new name at around this time (well, sort of, anyway). Atlas, the holiday home maker, employed Fred Braithwaite, previously with Cavalier tourers, to design the new luxury Hiltons: the 13ft Paris and the 15ft Marseilles. These smooth-looking, distinctive tourers proved too costly, though, and only around fifty were built before production stopped and Atlas returned to concentrating on holiday homes. Marquis, at Leyland Caravan Centre, had built 13 caravans over a period of 18 months. These were nice-looking vans and the clubman tourers were very well equipped.

In an effort to boost sales, ABI decided not to change its prices or models for the 1981 season. Thomson sales fell, and the company was finding that its once strong position in the caravan industry very much a thing of the past. Thomson was trying to hang onto the Glen series which was, by now, very long in the tooth.

Stephens and West, the maker of Stirling, Winchester and Cygnet, went out of business in December 1981. However, two former employees, Dick Harding and Tony Biggs, combined their 29 years of experience with Stephens and West and began a special Stirling, Cygnet and Winchester repair centre at Cricklade in Wiltshire. For a short time Stirlings could also be built to special order, although only the occasional special was produced.

It was around this time that Pemberton stopped producing Knowsleys at its Dovedale factory. Knowsley had, once again, fallen by the wayside (resurfacing yet again in 1986, this time built by Trophy).

It wasn't long before other manufacturers went out of business. Trio, Viking, Cabrera, Cheltenham, Dormobile, Fleetwind, Royale, Coronet, Windsor, Monolite (relaunched) Alpha Autosport and Portman all fell by the wayside within the first 18 months of the new decade. Cheltenham production may have stopped, but John and Tina Bradley took on the work of servicing and refurbishing Cheltenham caravans at Bransgore in the New Forest, where, 20 years on, they still keep the Cheltenham name alive.

The 500/4's interior, with its light oak woodwork, giving clubman appeal. The L-shaped kitchen boasted an electric console, all at a cost of £5563.

Swift's profile had hardly changed, even on this 1981 Danette. The company, though, through constant investment, expanded as most others contracted.

Swift, which had been busy investing in new machinery and its workforce, built up sales with the Corniche and Swift 'ettes. In an effort to change the shape, while retaining Swift identity, it launched the Cottingham for 1981, after doing extensive wind tunnel tests. It was also the first Swift built using sandwich construction, and sported a twin roof-mounted spoiler.

When Monolite came back for the third time it was with the Transcontinental, a super aerodynamic tourer complete with moulded glass-reinforced plastic front panel. The van rear came complete with its own roof spoiler with the model name moulded into it. GLT Monolite was the first to put a full rear roof spoiler on a UK touring caravan. Monolite was in many ways ahead of its time, although wasn't that strong on quality. Monolite claimed that its tourer could easily be towed by small cars, and also reckoned that the faster you towed it the more stable it was!

Mustang showed its new National tourer, which was entirely moulded in glass-reinforced plastic, at Earls Court in 1981. Mustang gave it a full specification, including a shower and hot and cold water. The rear end folded up, just as that of the Berkeley Messenger did in the 50s. The next showstopper was Bessacarr's Admiral, which was almost a design exercise by consultant Roger Simpson. Striking exterior colours in grey, blue and yellow, along with blue interior decor, didn't do the Admiral any favours.

It was ABI which turned heads at Earls Court that year – and kept them turned! Tycoon was the name given to ABI's silver bullet-styled tourer, which had a super-aerodynamic profile. With its silver-grey finish it looked very futuristic, and had excellent sales potential. Only 100 were built for 1982, but in 1983 the Award range adopted it. The instant success meant that the Award became a sellout tourer, setting new standards in tourer design. ABI exported a special twin axle Award to America for a short while. Swift retaliated with the 1984 Corniche; which employed smooth moulded glass-reinforced plastic front and rear ends.

CI, meanwhile, had launched a new pop-top Sprite for 1982. Called the Compact, Sam Alper had great hopes for this new venture, thinking it would help to boost sales. In Germany, CI Safari launched its own version of the Compact. The UK version cost £2195 for the 12ft version which weighed in at 10.6cwt.

Reg Dean took over the A-Line company after it hit financial problems. He opened up in smaller premises and named his new company Deanline. Dean produced some interesting designs under the Deanline banner up to 1985.

Just before the CI crash, the 1982 Sprite range included the Super Sprite, with features like double glazing, a fridge, a heater and an extractor fan. This is the interior of the 1982 Super Alpine.

In response to demand for better-equipped, lighter models, the Sprite Super, Piper Clubman, Monza Supreme, Rambler Equipe, Perle Custom and Lunar Dino were all produced (also because demand for the basic tourer was declining). Scottish manufacturer Thomson made a last-ditch attempt in 1982 to bring its tourers back from the brink. Interiors were more luxurious and came with a heater, a fridge, an oven and a fire extinguisher. Exterior was still basically Thomson, though, but with some bold decals added.

After four years of continual financial loss, Thomson was wound up. Only ten years previously its caravan division had been planning sales domination! Several ex-Thomson workers started up Forth Valley Manufacturing where, apart from doing repairs, they built the Futura tourer, an almost direct Thomson clone. Forth Valley went out of business, though, in January 1990.

After losing Thomson and Astral, it didn't seem possible that the UK's major producer would also be in for a fall. Sam Alper's CI empire had sold off Bluebird holiday homes in 1981. CI followed ABI's lead and cut prices to help its now much needed sales drive. It all came crashing down in December 1982: the giant was brought to its knees by over-production and models in need of replacement. Fairholme had already been dropped and Sprite, Eccles and Europa looked about to be consigned to the history books. Terry Reed from ABI went to see if ABI could buy the company. Amazingly, a rescue plan was put together and, with some backing, the CI company began to produce tourers again. Sam Alper left the industry for good at this point to concentrate on his vineyard and passion for sculpture.

In 1981, Monolite used a Metro to demonstrate how easy its Transcontinental tourers towed, although this picture shows the then new Escort hatchback. Note the rear spoiler – a common feature on tourers now.

From the old Bluebird Europe days, the Europa of the 80s was a new concept in design, sporting a more streamlined appearance.

The old 45 acre site was reduced to just 8! The workforce was cut and quality stepped-up in an effort to improve the new CI image. Production had only stopped over the Christmas break and most CI dealers, although shaken, stayed with the CI franchise. Many were glad they did because the medium-priced Esprit and the ultra modern Cosmos Explorer were produced in 1984, both with impressive specifications. CI had made a comeback and went on to expand its workforce and increase sales.

Springfield Caravans at St Helens, known for making mobile work units, launched the medium-priced Acapulco, Miami and Manhattan range of tourers for 1982. Magnum, super clubman tourer manufacturer, newly based at Corby, and Essex repairer, Sherringham, both turned to caravan manufacture in 1984. Even in these lean times new tourer manufacturers were appearing.

New models by established makers were also being introduced. Bailey went for the budget market again in 1983 with the launch of the plain flat-roofed Caribou. Not since the Prima, in 1979, had Bailey looked at this sector. Elddis launched the Breeze and Compass, its Echo range. Even Fleetwood joined in with the lightweight Crystal. To celebrate the twenty-first birthday of Terry Reed's Ace brand name, in 1983, the Ace limited edition Debutante was launched, with some little extras. The following year Ace launched the Marauder, basing the shell and interior on the Debutante.

The Birmingham NEC staged its first caravan show and witnessed the launch of several new models. The NEC was to expand as a caravan and boat show, eventually outgrowing Earls Court. More shows were organised, one being the major Scottish show at Kelvin Hall Glasgow. By 1985, after a few hard years, things seemed to be on the up (the 1984 season had even seen shortages of some new tourers).

New model ranges were introduced and features, such as mains electric and double glazing, became standard on the main mid-priced models. Gas lights were a thing of the past as well. The end kitchen, after many years, began to give way to the new rear end washroom design. ABI, Avondale and Leda were some of those to use this relatively novel layout.

Glass-reinforced plastic was now being used as commonly as prefinished aluminium had been in previous years. Swift's Cottingham was replaced with one of the most popular tourers for years. The Challenger became a sell-

ABI Awards from 1986 closely followed the lines of the original 83 model year tourer. ABI sold the Awards in Holland as well as in the UK. From day one there was always demand for these aerodynamic tourers.

out model range with its modern interiors and super slippery shape. An integral front locker had become a must with the advent of moulded panels. Even restrained clubman maker Castleton now used new glass-reinforced plastic mouldings for roof, front and rear panels. Elddis, Lunar, Leda, Viscount, Island Romini and the new futuristic Elan had all gone over to super aerodynamic styling, setting the pace for the 1990s.

Bailey, who had been taken over by the Howard family from Winn Industries in the 1970s, went from a manufacturer of respected, middle-of-the-road tourers, to the extremely successful forward-thinking company it is today. The first real signs of this were in the mid–80s, with the Pageant range, which included a shower tray and even a spare wheel as standard. Mardon, with its Meridian and Mystique models, was not far behind. The Mystique replaced the deleted Sovereign. However, the death of founder Les Marshall meant that his son, Peter, had to take over. Mardon carried on for a number of years but with only limited success.

Over in Cottingham, Swift gave the 'ette model range a complete makeover, further increasing sales and also giving the Swift ranges a corporate image. Trophy took over the Knowsley name in 1986 and tried to give it a new lease of life (basically a Trophy in all but name and pitched just above the budget Medallions). The range failed to inspire new or old Knowsley followers and the name disappeared for good. Towards the end of the 1980s, two very significant names within the industry arrived: Coachman and Vanroyce.

Vanroyce was formed by Tommy Green

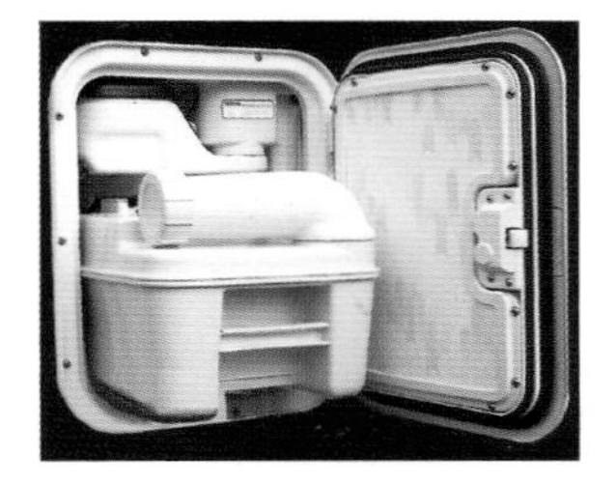

and John Darwin (joined later by twin brother Cyril). These clubman tourers soon found themselves a niche, beating even the smart-looking Cotswold Celeste and Windrush models. Vanroyce offered a distinctive tourer at a reasonable price.

Coachman, over in Hull, was a new manufacturer started by George Kemp and Jim Hibbs (both ex-ABI directors) and Clive Bradfield (ex-Lynton sales manager). Lynton had gone out of business in February 1986 and Coachman provided well equipped tourers which competed keenly with ABI.

Two manufacturers that didn't quite establish themselves as successfully were BPH Designs, maker of Buzzard super luxury custom-built tourers, established in late 1983, and Andromeda, over in Goole. Danny Galloway produced his little Andromeda Mayfair tourers in several layouts. Buzzard may have had a very low profile, but still continues at the time of writing, producing special clubman tourers to order.

After the demise of Mustang in 1984 the name looked set to bounce back, with motorhome maker, Acorn Conversions, using old Mardon panels. Just as with Knowsley, though, the Mustang didn't make any impact.

With the late 1980s came the "trendy" tourer. Abbey added "go-faster body stripes" to SR models with their red-dominated interiors,

The1986 Elddis Shamal GT was another winning design from the ABI stable. The GT range was well equipped for its time.

Swift's response to ABI was the smooth-looking Challenger range. These sold well in the UK, with some also going for export. This 1987 440/4 still looks good 12 years on.

Challenger interiors of 1987 were up-to-date and U-shaped front seating was a strong selling feature. This was possibly one of Swift's best-selling ranges.

and ABI did an equivalent interior makeover, based on the Monza, and called it the Mode. ABI also launched the Monza-based Disco, which, again, had a predominantly red-striped interior! Even Avondale had a go with its four van Mayfair range. Yellow and greys were the Avondale designers' choice of colours.

Abbey then went one step further with the Lifestyle, a van designed by an industrial designer. Although listed, only one model was built. This trend, or fad, didn't last more than two years, though, (it seems UK buyers were too conservative).

It is worth noting at this point that exports were increasing, with ABI being the main exporter. Caravans International was also picking up new orders, and both ABI and CI were recognised with Queen's Awards for exports.

Lynton, which had split from its special units division a few years earlier, had invested heavily in the Excalibur, a wheel at each corner tourer. Based on its VIP range, this big family tourer met with a poor response. Unlike the earlier budget Sport models, which Lynton took over from a failed local project, the Excalibur was a disaster and the company went out of business in February 1986.

In the Midlands, just round the corner from

The first 1987 Vanroyce model, launched at the Hull Gibson Lane trade show, became an overnight success in the clubman sector. Tommy Green and John and Cyril Darwin went on to form Vanmaster in 1995.

Cotswold eventually became a member of the Swift Group but, in 1988, was still independent. A superb classic tourer, with stacks of standard equipment, this is the Windrush 132.

Avondale, John and Dawn Owen were building their Centaur range. The two models, the Corfu and the Taura, both 13ft two and four berth models, looked very much like Avondale's Perles and were made at the rate of one a week. The Owens moved the company over to Wales, into a new unit near Lake Bala, repairing as well as selling accessories. In 1990, the company folded, sales had been on a direct basis only, but dealers were being sought before the company's demise. Rovahome, a small specialist maker in Norwich, produced good-looking glass-reinforced plastic tourers from 1985. Rare when still being made, they became even rarer when production ceased in 1990.

Swift had now moved into motorhome production, along with Elddis, Compass and Lunar. Swift took the expansion stage further, by buying Cotswold Coachcraft in 1989. Swift hadn't intended to shut the Grimsby factory, and ended up moving it over to Dunswell Road. Cotswold, like Buccaneer, Castleton and Carlight, had used the old coachbuilding construction method, but Swift soon changed this to bonded construction.

Just before the Swift buy-out, Cotswold introduced two super luxury clubman vans named Spirit. Bristling with extras, such as an on-board vacuum cleaner, the limited edition (35 were built) was available only through Lancashire dealers W&A Hartley, priced at £13,550 to £14,950. These were soon dropped, though.

Carlight had implemented a policy of selling direct, and even took in used models, reconditioning them before selling them on again to those that couldn't afford the £17,000 price tag of a new one.

Business was booming again as the 80s came to an end. Quality became more of a byword for manufacturers and Bailey and Swift ploughed thousands of pounds back into quality programs. Swift took on a new designer in 1988, Richard Emerich. Emerich came from specialist vehicle builder, Coleman Milne. Ten years later he moved to the newly formed ABI UK as chief designer. Swift had also purchased more land and expanded its empire further.

Compass and Lunar expanded their factories to keep up with demand. Lunar's two berth Meteorite and Compass's Omega and Rallye ranges were in constant demand. Lunar joined the budget market with its super lightweight Micron, taking on the likes of ABI's Marauder. Swift went budget with the Rapide 12/2 and 12/4, basing them on the old Swift shape and interiors. At over £3700, they proved extremely popular, with new layouts being added. Elddis, too, latched on to this market, producing its Wisp special (1988 saw it become a definite model range, joining the XL and GTX ranges).

Bailey, which was becoming more popular with models like the Chieftain, Senator and Scorpio, was seriously into quality, with a department that tested build and component durability. The Howard family's determination to succeed meant Bailey picked up new dealers and design awards. This low-profile Bristol company became the high flyer of the 90s.

Meanwhile, CI had a showstopper at Earls Court in 1987 with the rather sci-fi-looking

Voyager. Steve Trossell, previously with Bessacarr, designed the interior which combined blue soft furnishings and ash wood finished furniture. Brimming with loads of advanced features, such as solar panels and a hydraulically operated corner steadies, the Voyager remained only two years before being deleted.

For 1989, Swift responded with a new windcheating Corniche, which came with full glass-reinforced plastic moulded panels, making it a very aerodynamic tourer. Interior-wise, Swift played safe and kept it traditional and well specified. The Corniche was a big success.

Avondale also produced an all-new shape for its clubman Avondales. Like Swift, Avondale used computer-aided design to come up with the shape.

With the end of the 80s, the industry experienced increased production levels and equipment such as blown air, flyscreens and blinds, hot water, mains electrics and better insulation - all became standard.

The increase in dealer specials had begun, too, ABI was the main manufacturer in this expanding market, its specials usually being high specification Marauders. Examples included Harrington's Celebration, Barron's and Quality Caravan's Quasar, Cross Country's Cricket, Yorkshire Caravans' Yorkshire Rose and Burtree's Starlet. Elddis also became seriously involved in this area, with its Vogue models being built for Barrons. In May 1993, *Caravan Magazine* celebrated its 60th year, and the magazine's long-standing editor, Barry Williams, commissioned Swift to build 100 special limited edition Diamond Corvettes with *the Caravan Magazine* logo.

The Robin Argyll was based very much on the old Astral Cameo from 1980. Even the chassis came from Astral when that company was shut down.

The new decade saw ABI go for single dealerships, forcing established ABI dealers to give up the franchise or, as in some cases, open a new sales centre or take over an existing one. After a few years it was clear that this policy hadn't worked, nor had the fixed-price campaign, so ABI went back to multi-franchise dealerships. Tony Hailey, who had been with Cosalt for his caravan industry years, joined the Swift group, further strengthening management there. As Hailey joined, David Rochester, who had been with Swift since the early 70s, left to join Lunar as sales manager. Elddis had a management buy-out and cut connections with ABI.

ABI also decided to stop introducing new models on a yearly basis, doing so only as and when required. This policy failed, however, and one reason was that it confused buyers, so ABI went back as before. CI Sprites, Eccles and Esprit models had maintained market share, and export sales levels, but trouble on the horizon meant the company became involved in a management buy-out. This secured the jobs of the 300 strong Newmarket workforce, for the time being at least. The new company, Sprite Leisure, strengthened the model range with the rebirth of the Europa range for 1994. It was an instant success and featured a choice of two interiors (called the Classic), traditional furniture and soft furnishings, or furniture with a grey wood finish. Sprite Leisure clearly had a winner.

Silverline, which made a comeback with its narrow, tiny, luxury tourers, had run into financial problems, and, by the end of 1991, the Silverline name was laid to rest. Mardon, which had improved its ranges in the late 80s with strict quality control just like Bailey, was the first to introduce a long-term, three-year warranty. By April 1991, the parent group, Marshall Holdings, went into liquidation. Several months elapsed before the new Mardon range was shown for 1992 at Hull trade shows. Now owned by the Hallmark company, which produced holiday homes as well, the new Saffron, Orchid and Orchid GT models came from the drawing board of John Swift, formerly with the Swift Group.

The main concern was to get old dealers back and also enlist new ones. Mardon was briefly back in business (before going under yet again). Two former employees, Colin Stevenson and Janet Millington, began a dedicated Mardon products repair business in 1992. Riding

Caravans, as they became known, began a four model line-up. For 1994, the Silverstar name was rekindled but materials stolen from the factory meant production stopped yet again.

Cosalt's Abbey, and the reborn Piper Caravans, watched tourer sales take a dive. Profits were down and the GT and GTS tourers which had proved so popular couldn't turn the company's fortunes around. By the end of 1992, Cosalt decided to sell off its touring caravan division. Swift, waiting in the wings, saw the opportunity to purchase one of the UK's top brands. Abbey Caravans moved to Cottingham, and the old Grimsby factory was shut down. Swift learned from its mistakes with the Cotswold brand (which hadn't worked out; no models made after 1991), and developed the Abbey range, continuing and increasing sales.

From the narrow-widthed Trek Iona to the super twin axle Spectrums, Abbey set new standards. At Earls Court in 1995, the company produced its most controversial tourer to date, the Abbey Domino. This 11ft two berth tourer had a new maple style furniture design, but main talking points were the glass-reinforced plastic sides, an idea taken from the Swift Motorhome division. It met with some success and so the Domino four berth, end bathroom, 17ft model was launched for 1997, quickly followed by the Chess and Solitaire. Abbey called them the Evolution range – as yet no other maker has followed Abbey's lead. For a limited period, part of the old Abbey factory at Grimsby became active again.

Some of Abbey's former employees began a new business called Custom Caravans. John Siddal, the man mainly responsible, also began making custom-built caravans and even houseboats! Astra touring caravans, though, became the small firm's main interest. Building 12 to 15ft tourers, these vans showed promise with quality workmanship and practical designs and layouts. Launched in 1994, the company signed up several dealers, and even, at one point, built a few export models for a holiday home manufacturer, as well as some dealer specials. The Cosalt-owned factory was sold off, leaving the Astra manufacturer with no affordable premises, and so another caravan maker became consigned to the history books.

When the old Cotswold company was taken under Swift's wing, several of the original workforce started up a new company, Shire Coachcraft, in 1991 and produced the Royalty range. The design of these traditional tourers relied very heavily on the original Cotswold

The Compass Omega from 1987 was a popular tourer from the ex-Elddis founder. Now, ironically, it is made by the Explorer Group-Elddis!

tourer. Bob Clarke, one of the company's founders, had been the buyer at Cotswold. Although craftsmanship was high and standard equipment good, production at the Stallingborough-based plant came to a halt after a few years.

Lancashire-based Trophy had financial troubles and, as the company came close to collapse, local motorhome manufacturer Venture bought out Trophy, renaming it Trophy Leisure. Venture had a short run of success with the Ascot and Cheltenham ranges, before being sold off to a haulage firm in Preston. By 1995, however, the Trophy name had gone.

Elddis Caravans took the decision to produce a budget tourer and launched the Breeze at GMEX. This was basically a rehashed range (essentially a Wisp) which was shortlived. Two years later, in February 1994, the Crown range resurfaced at the NEC Show. ABI had left the name with Elddis and, as the tourers were built at the Elddis factory, they were marketed as a separate brand. The Crown and Golden Crown were basic budget tourers but with a degree of comfort. Coachman introduced the continentally-influenced Concept in 1996, again at the NEC, but it was a failure.

Castleton, whose tourers had rather old-fashioned interiors and limited clubman appeal, introduced the Genesis. Using bonded construction and ash wood furniture finish, it was a totally new concept at Castleton but it didn't save the company and Castleton went out of business after nearly 40 years.

Elddis sprang a surprise at Earls Court in 1996. It had nicked the Genesis name and put

Bessacarr's 1983 range had the Cameos as the company's move towards twin axle tourers. Bessacarr did a great job in making a success of the defunct Astral Cameo.

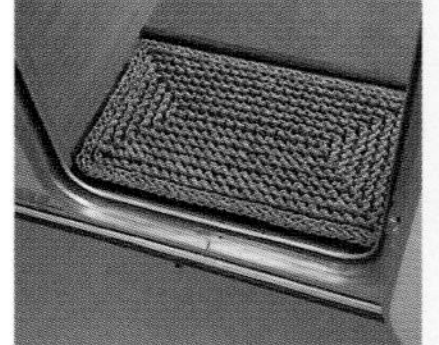

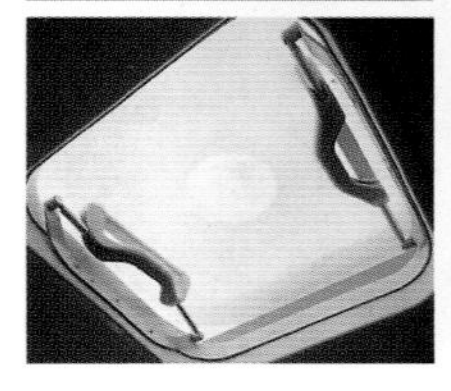

it on its own van, which had a very futuristic design (glass-reinforced plastic panels played a big part in it). Its windcheating shape drew much attention and it boasted a built-in jacking system. The interior had all the features you could need: furniture included wicker-finished cupboard doors, combined with striped regency upholstery. The Genesis was dropped for 98 but the shape was used in a toned-down version for the GTX and Crusader ranges, which were added to the XL for 1999.

After Silverline's demise, Edwin Robinson was assigned to watch over large Essex dealer Upminster Caravans' venture into caravan manufacture. Upminster had several branches, and had been a large ABI dealer. Eddie Hatter, the owner, had a new factory built at Annie Reed Road at Beverley, to build his new tourer range (Craftsman Caravans was the new name), which he would sell through his own dealer outlets and also from the factory gates. Hatter sold his vans direct at factory prices, offering incredible value for money. The Beverley was the budget van, followed by the mid-market Miracle (the best seller) and the Superior range. A Craftsman sold for around £1000 less than the competition. Not only did this cause uproar, but also the fact that the Craftsman profile and interior were an almost direct copy of Elddis models. Hatter continued to sell his vans, and some even went to the Dutch market. Craftsman's prices remained the same for two consecutive years, though, by 1995, Craftsman tourers had stopped and Hatter returned to tourer retail.

Vanroyce, whose distinctive luxury tourers were almost an icon of the clubman van, had won numerous design awards. The company moved to a new factory at Skelmersdale, becoming part of the Mosley group. At this factory was produced a cheaper Vanroyce, the Landseer, which had limited success. Sales, in fact, were slowing down, and, by the end of the 94 season, Vanroyce stopped production. ABI stepped in at the eleventh hour and bought Vanroyce, moving production to the company's Autotrail motorhome factory at Immingham, where, under Stuart Turpin, the Vanroyce tourer went back into production.

Tommy Green and John and Cyril Darwin, who had left Vanroyce just as the crash came, soon started up a new venture, Vanmaster, in late 1994 (the name was taken from the original company after John Darwin saw one of the old Vanmaster tourers). By the 1996 model year, the new Vanmaster super luxury clubman tourer was built. This was a brave move, given that Castleton had gone into receivership, closely followed by Bessacarr. Vanmaster established itself quickly, with a reputation for detail, quality and innovation.

A Coachman 460/4 from 1988. Coachman tourers soon became a familiar sight on the roads, after the initial launch at Birmingham's NEC Show in 1987. By 1998, the company had been swallowed up by the Elddis Group.

In the meantime, Swift had bought the Sprite Leisure group in September 1994, though, by the following year had dropped name Sprite Leisure and relaunched the company as Sterling Caravans. Alpine, Musketeer and Major, all well established names, were dropped in favour of numbers. Swift then made the decision to move all 1997 production to Cottingham. Newmarket's days as a site of major caravan manufacture were over. The site has since become a business park, the only connection with caravans being the Newmarket Caravan retail centre headed by ex-CI man Dave King.

In the mid-90s the Sprite was dropped, though it was still produced for Swift's export markets. The Europa and Eccles brands were further improved and have since become some of the Swift Group's best sellers. After the

An ABI Jubilee Rallyman from 1989 proved a popular choice for the seasoned caravanner. Nine years later when ABI Ltd crashed, this innovator resurfaced under the ABI UK banner.

Craftsman's Miracle range was the company's short-lived best-seller. Its Elddis looks, and competitive price, gave the company brisk sales in the early 90s.

Cotswold experience, Swift still needed a super clubman tourer, and found an excellent opportunity for a ready-made market when Bessacarr's parent company, the Arnold Lever Group, wanted to dispose of its caravan connections.

Bailey, which had taken more of the market, launched its reasonably priced, well-equipped Beachcomber range in 1993.

Ironically, 1996 was definitely Bessacarr's best model year, with quality finished glass-reinforced plastic mouldings and high quality cherry woodwork. Swift, learning by its Cotswold mistakes, invested heavily in the Bessacarr to create a better unit than before, and it worked - up to a point. Swift was now in direct competition with Vanroyce, Vanmaster and Buccaneer. The latter used a full glass-reinforced plastic roof and new front and rear panels, which brought the Buccaneer range up to date. Natural oak veneers and luxury soft furnishings, as well as established models such

Swift Group designer, Andy Spacey, took on development of the 1996 Abbey Domino. Glass-reinforced plastic sides are said to be more practical than aluminium panels. For 1999, three models exist, and, as yet, no other maker has gone down this path.

as the Elan and the Schooner, gave the company a loyal following, with its traditional coachbuilt techniques.

A big talking point for Avondale would be the export Panther range introduced for the UK market and renamed the Landranger. Designed with 4x4 vehicles in mind, these large tourers provided full living space and 300 were sold the first year. Avondale also introduced its Dart range in 1998, dropping the long-standing Perles. In 1999 the Leda was replaced by the Rialto, which was highly rated by the caravan press.

At the NEC in 1996, Bailey introduced its icon of the budget tourer world – the Ranger. Priced at a very competitive £8500, the Ranger came with mains lighting, a full oven, flyscreens and blinds, plus blown heating and a TV aerial. It was an overnight success, with newcomers and experienced caravanners alike buying them. Bailey didn't look back; in fact, the Bristol-based firm sold over 1000 Rangers in the first twelve months. Actually, all Baileys were selling exceptionally well, and the company even started exporting to Holland (something which hadn't been done for 17 years. For 1998, Bailey further strengthened the range by adding the Hunterlite, which was priced at just below the Ranger.

Lunar, which had produced some good tourers, such as the Delta, Planet, Clubman and Meteorite, was experiencing tough competition from the likes of Bailey's Ranger, Abbey's Iona and the Compass Lynx. The Ariva, LX2000, Lexon, and the budget Solar, were all introduced by Lunar to combat the opposition. The narrow-width Ariva, introduced in place of the Meteorite, was even exported to Japan. The LX2000 and Solar proved very popular, and, due to their lightweight design, were more practical for the smaller car.

More recently, new manufacturers are rare, but two brothers, John and Paul Mcdonald, left Lunar to start a caravan repair business. In the last couple of years they have ended up building tiny tourers called JNC, finding a niche market for their value-for-money lightweight tourers. Given the right conditions, JNC may be a new name to watch out for.

As we come right up to date, it is time to take a look at the disaster that befell ABI. ABI was building tourers for France, Germany and Holland, and so the company's exports and general state of health were good (it was the third largest company in Europe and the biggest in the UK). Its famous Awards, Jubilees, Aces and Sprinters (which replaced the Monza) meant that ABI models were most caravanners' first choice. ABI had given us the Ibiza, a continental UK specification tourer, plus its last controversial tourer, the budget rear-doored Adventurer. Things weren't as good as they seemed, though, and in 1998 the company was in great financial difficulty, so much so that 600 workers lost their jobs, and the company was brought to its knees.

ABI crashed, just like CI, which left Swift as the largest manufacturer. Elddis bought out Coachman, along with Compass, which took it

The ABI Ibiza was a flop, its trendy graphics and interior failed to impress all but a few buyers.

In 1996 Bailey created a storm with the Ranger. An instant sales success, the model offered high standards at an incredible price - a market leader in its class.

Swift's ultra classy Corniche takes the range into the millennium with aerodynamic style.

Lunar's 1999 LX2000 replaced the longstanding Planet range, and has proved a sales success for this 30 year old company. Hard to believe Lunar began in 1969 in an old barn.

Ironically, the classic Cheltenham still carries on, with the help of the Bradleys down in the New Forest. Cheltenhams are restored there, helping to keep this classic design going into the next century.

In 1999 Compass offered the Kensington as a competitor to the Ranger. The 500/4 costs £10,495 and has a full specification.

to second place. For 1998, Compass brought out two Herald ranges, the Emblem and the Clarion, both aimed at Bailey's Rangers, and the Elddis Group decide to give all the brands a new identity. The Explorer Group, which comprised Elddis, Crown, Coachman and Herald, plus the motorhome division, was to be the new force in the industry.

Just as time was running out for ABI, an eleventh hour deal saved the day and it became ABI UK Ltd. In 1999 it made a splash with new and revitalised ranges like the Quartz, the Manhattan and the Ikea-styled Papillon range. Although not the same as before the crash, ABI has tried to inspire confidence in its dealers and customers, and is building exports. But the company is far smaller, in terms of plant size and position within the UK caravan industry, than it was before.

The interior of the 460/2 Kensington shows the up-market decor and now very fashionable end bathroom.

A year or so later, however, ABI decided to stop tourer production altogether and concentrate instead on its highly successful holiday home caravan range. The Ace name, however, has come back, this time under the Swift Group which has relaunched the Award and Jubilee ranges.

Fleetwood, in its Suffolk-based factory, had always been on the industry sidelines, never making any great statements. Fleetwood's policies are now changing, as, in the last few years, the company has brought out some interesting models, such as the 1999 Heritage range, a touring caravan with plenty of "wow' factor.

Richard Emerich moved from ABI UK and went to Fleetwood (now part of Adria) further developing the Fleetwood brand of tourer.

The Explorer Group now comprises Compass, Elddis and Buccaneer (having acquired the Buccaneer Company with Michael Hold in charge of design and marketing). Explorer sold off the Coachman Caravans division, leaving the Hull-based company back as an independent maker.

The British caravan industry has had its fair share of highs and lows, but, despite the problems, always seems to recover. Although large concerns are good for the industry, with respect to financing tourer development and caravanning as a whole, what the industry also needs is individual concerns with new ideas to keep customer choice as wide as possible.

The British touring caravan has always been innovative, and, as we go into the millennium, the UK industry will maintain this enviable reputation. I firmly believe that the industry still lacks positive support, especially from the media in general, but hopefully my two books will help show how important this British industry is. The UK touring caravan has certainly come a long way since those early pioneering days.

Index